Plant Science - Structure, Anatomy and Physiology in Plants Cultured *in Vivo* and *in Vitro*

Edited by Ana Gonzalez, María Rodriguez and Nihal Gören Sağlam

Published in London, United Kingdom

IntechOpen

Supporting open minds since 2005

Plant Science - Structure, Anatomy and Physiology in Plants Cultured in Vivo and in Vitro
http://dx.doi.org/10.5772/intechopen.78878
Edited by Ana Gonzalez, María Rodriguez and Nihal Gören Sağlam

Contributors
Wen-Lii Huang, Hsiang-Ting Lee, Masaaki Watahiki, Dongyang Xu, Shivani Lalotra, A. Hemantaranjan, Bhudeo Rana Yashu, Rupanshee Srivastava, Sandeep Kumar, Marcelo Pace, Admasu Moges, Yohannes Moges, Elizabeth Fontes, Miloud Hafsi, Ali Guendouz

First published in London, United Kingdom, 2020 by IntechOpen
IntechOpen is the global imprint of INTECHOPEN LIMITED, registered in England and Wales, registration number: 11086078, 7th floor, 10 Lower Thames Street, London, EC3R 6AF, United Kingdom
Printed in Croatia

British Library Cataloguing-in-Publication Data
A catalogue record for this book is available from the British Library

Additional hard and PDF copies can be obtained from orders@intechopen.com

Plant Science - Structure, Anatomy and Physiology in Plants Cultured in Vivo and in Vitro
Edited by Ana Gonzalez, María Rodriguez and Nihal Gören Sağlam
p. cm.
Print ISBN 978-1-78984-746-8
Online ISBN 978-1-78984-747-5
eBook (PDF) ISBN 978-1-78985-562-3

For EU product safety concerns:
IN TECH d.o.o., Prolaz Marije Krucifikse Kozulić 3, 51000 Rijeka, Croatia, info@intechopen.com or visit our website at intechopen.com.

Meet the editors

Ana Maria Gonzalez is the director of the Northeast Botanical Institute, where she is an independent researcher in the Career of Scientific Researcher of CONICET. She is also in charge of the Plant Anatomy Laboratory at the Institute. Her teaching career includes being a Professor of Biology at the Universidad Nacional del Chaco Austral (Saenz Peña, Chaco, Argentina), and of Vascular Plant Morphology, Faculty of Agriculture. National University of the Northeast (Corrientes, Argentina). She is a postgraduate professor in the master's and doctorate courses of the Cs. Agrarias Faculty, with numerous courses on anatomy and histological and bio-imaging techniques. She has seventy-five published works and more than 120 papers presented at national congresses and seventeen in foreign countries.

María Victoria Rodriguez currently works at the Facultad de Ciencias Bioquímicas y Farmaceúticas (FBIOYF), Rosario National University. She is an assistant researcher at the Consejo Nacional de Investigaciones Científicas y Técnicas (CONICET) of Argentina. She is an adjunct professor of Pharmacobotany for Pharmaceuticals students and Cell Differentiation and Structure of Vascular Plants for bachelor students in Biotechnology. Dr. Rodriguez has presented more than sixty works at scientific-technological events and has twenty-eight publications on topics related to plant anatomy, pest biocontrol, bioinformatics, and botany (medicinal plants). Dr. Rodriguez has twenty-five postgraduate courses and/or trainings on the subjects of her specialization. She is a reviewer of international journals and book chapters related to plant anatomy, medicinal plants, and biocontrol.

Nihal Gören Sağlam obtained a Bsc (1995–2000), Msc (2001–2004), and PhD (2004–2009) in Plant Physiology from Istanbul University's Department of Biology. From 2002 to 2009 she was a research assistant at Istanbul University, Department of Botany. From 2010 to 2011 she was a post-doctoral researcher at Warwick University, School of Life Sciences, UK. Since 2009 she has worked as a research assistant doctor at Istanbul University in the Department of Botany. Her research experience/interests include plant physiology, senescence hormone interactions, stress senescence interactions, microarrays, gene expression, protein–DNA interactions (Yeast one-hybrid), and confocal microscopy. She has had training as a post-doctoral researcher (2010–2011) at the School of Life Sciences, Warwick University, UK, (Tubitak Grant); an academic visitor (2008; 6 months) at Warwick HRI, UK; "Molecular Aspects of Salt and Drought Tolerance in Crops (Plants)" (2007) at the University of Cairo, Giza, Egypt (ICGEB Grant); and an academic visitor (2006; 3 months) at Warwick HRI, UK.

Contents

Preface

This book presents comprehensive information on plant anatomy, physiology, and morphogenesis as well as the conditions of the different environments to which plants are exposed. It highlights the importance of knowledge of the anatomy of plant tissues for different applications. In addition to the variety of physiological studies presented here, the book also emphasizes anatomical studies in botanical quality control of medicinal herbs with human health benefits.

Chapter 1, by Marcelo Pace (Mexico), describes in detail the origin and structure of phloem and its cell types. The author, in addition to analyzing the composition of phloem in different groups of plants, highlights some of the known commercial uses of this tissue.

Several contributions deal with the relationship between various hormones with processes in plants, both ex vitro and in vitro, and under different conditions of stress. The chapter by Dongyang Xu (China) and Masaaki Watahiki (Japan) introduces the development of root systems regulated by various phytohormones like auxin, cytokinin, and others. In their chapter, Hsiang-Ting and Wen-Lii, Huang (Taiwan) study the relationship between endogenous phytohormone signals and carbohydrate metabolism during regenerable callus induction induced by osmotic stress in rice. Based on the prolific role of jasmonates and their derivatives in different fields of biological sciences, Shivani Lalotra, A. Hemantaranjan, B. R. Yashu, R. Srivastava, and S. Kumar (India), in their chapter, discuss these phytohormones in terms of future agricultural, biotechnological, and physiological research. The onset of leaf senescence is a highly regulated developmental program that is controlled by both genetics and the environment. Otto Teixeira Fraga, B. Paes de Melo, L. Fernando de Camargos, D. Pellanda Fagundes, C. Cabral Oliveira, E. Bassi Simoni, Pedro A. Braga dos Reis, and E- Pacheco Batista Fontes (Brazil) analyze a regulatory circuit integrating stress-induced with natural leaf senescence.

Hafsi M. and Guendouz A. (Algeria) evaluate the performance of some durum wheat genotypes and test the efficiency of using senescence parameters in screening under semi-arid conditions.

Admasu Moges and Yohannes Moges (Ethiopia) present a review that documents medicinal plants used for traditional treatments as well as their parts, usage, ecology, and quality control. Accordingly, they review eighty medicinal plant species whose leaves and roots are the main parts used for preparation of traditional medicines. Chromatography, electrophoretic, macroscopic, and microscopic techniques, and pharmaceutical practice, are mainly used for quality control of herbal medicines.

Ana Maria Gonzalez
Academic Editor,
Instituto de Botánica del Nordeste. UNNE-CONICET,
Argentina

Maria Vitoria Rodriguez
Universidad Nacional de Rosario,
Argentina

Nihal Gören Sağlam
Istanbul University Department of Botany,
Turkey

Chapter 1

Phloem: Cell Types, Structure, and Commercial Uses

Marcelo R. Pace

Abstract

Phloem is the vascular tissue in charge of transport and distribution of the organic nutrients. The phloem is also a pathway to signaling molecules and has a structural function in the plant body. It is typically composed of three cell types: sieve elements, parenchyma, and sclerenchyma. The sieve elements have the main function of transport and typically have lost their nuclei and other organelles in the course of their specialization. Hence, the sieve elements rely on specialized neighboring parenchyma cells to sustain all of their physiological function and activities. All cell types of the phloem may vary morphologically and in their distribution in the tissue, and this diversity is taxonomic and functionally informative. The phloem can be of primary or secondary origin, being derived from either procambium or cambium, respectively. Some vascular plant lineages have exclusive primary phloem, such as the lycophytes, ferns, and the monocotyledons, and the sieve elements will be long living in these taxa. In plants with secondary growth, the secondary phloem is formed, and typically the primary phloem collapses. Because new secondary phloem is constantly formed, the longevity of sieve elements in the secondary plant body is much more reduced. In this chapter, the structure of the phloem and its cell types are described in detail and also some of the known commercial uses of this tissue.

Keywords: sieve tube, sieve tube element, companion cells, bark

1. Introduction

Phloem is the vascular plant tissue responsible for the transport and distribution of sugars produced by the photosynthesis. Since the plant is a continuum, phloem will be found in the external part of root cylinders (**Figure 1a**), in the stem vascular bundles (**Figure 1b**) and in the abaxial part of the venations of every single leaf (**Figure 1c**). While the most common is to have the phloem external to the xylem in roots and stems and abaxial in leaves, some exceptions exist and are usually taxon specific. The phloem found in the inside is named internal or intraxylary phloem (**Figure 1b**).

As a constitutive tissue in the plant body, phloem functions extrapolate its main function of sugar transport, including transport of signalizing molecules such as mRNAs, hormones, defenses from biotic and abiotic agents, sustenance of the organs, gas exchange, and storage of many ergastic materials, such as starch, calcium oxalate crystals, and tannins. Parenchymatic cells of the phloem can also give rise to new meristems, such as the phellogen or cork cambium. All vascular plants have phloem, which typically includes specialized living conducting cells

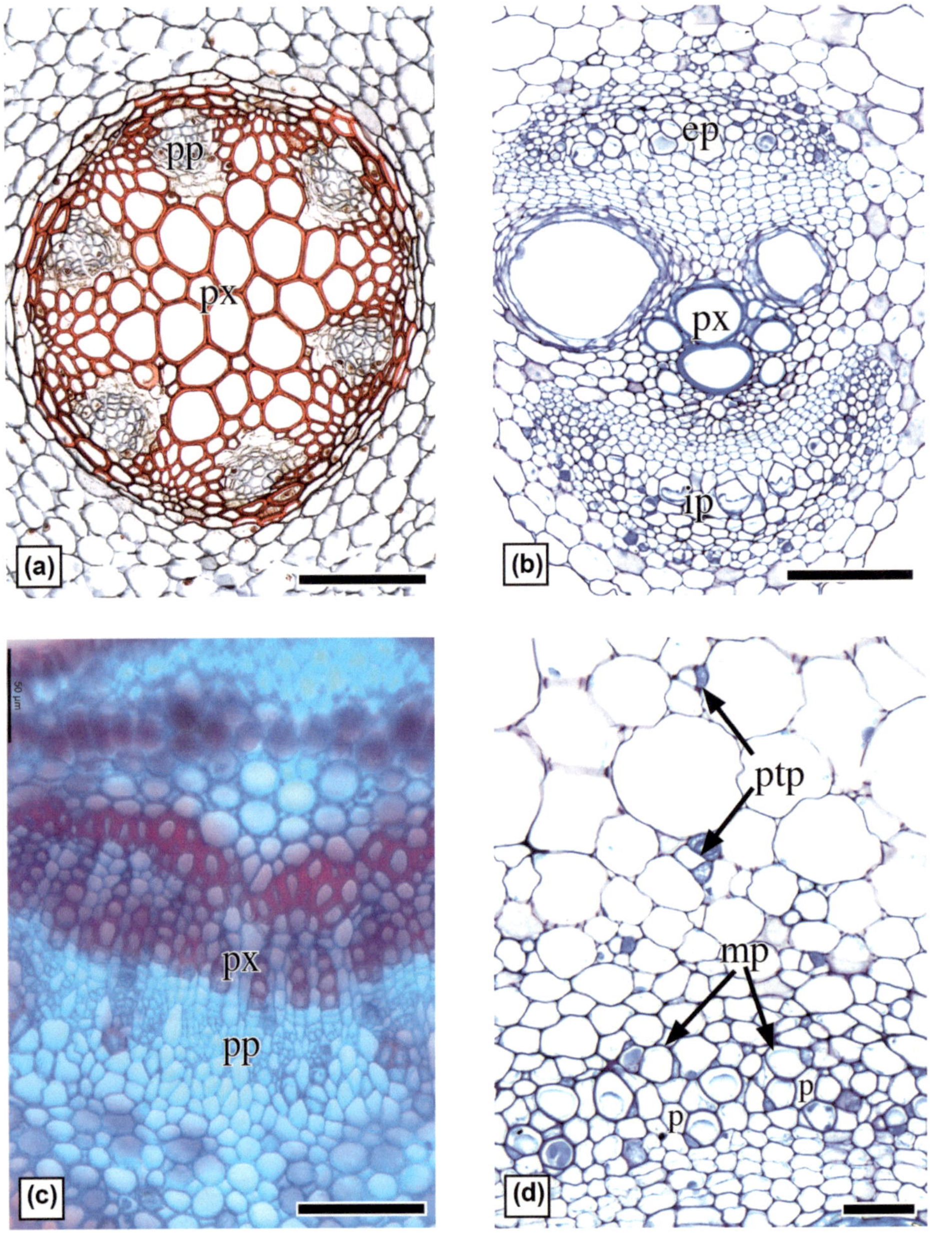

Figure 1.
Location of the primary phloem in different organs and its cell composition. (a) Ranunculus acris (Ranunculaceae). Root transverse section (TS), exarch structure, six strands of primary phloem alternating with the six protoxylem poles. (b) Bicollateral vascular bundle of a squash, Cucurbita pepo (Cucurbitaceae) TS. On the top is the external phloem, and on the bottom is the intraxylary or internal phloem. (Picture credit to Solange Mazzoni Viveiros). (c) Detail of the leaf midrib vascular cylinder of Tetrapteryს mucronata (Malpighiaceae) showing primary xylem on the top and primary phloem on the bottom. (Picture credit to Leyde Nayane Nunes). (d) Detail of (b), showing the protophloem on top and the metaxylem on the bottom. ep, external phloem; ip, intraxylary phloem; mp, metaphloem; p, parenchyma cell; pp, primary phloem; ptp, protophloem; px, primary xylem. Scalebars: a, c, d = 50 µm, b = 130 µm.

named sieve elements whose nucleus, ribosomes, and other organelles degenerate during maturation, making sugar transport more efficient. The life and function of these cells will then rely on closely associated parenchyma cells which support the physiological functions of these sieve elements [1]. Although typical phloem is exclusive of vascular plants, rudimentary phloem-like conducting cells are present also in other lineages, such as the bryophyte leptoids, and even outside the plant kingdom, as the trumpet cells of the kelps and phaeophycean algae [2]. The primary phloem derives from the embryo and the apical meristem procambium throughout the life of the plant or from the cambium, in plants with secondary growth.

2. Phloem cell types

The phloem is a complex tissue and is formed typically by three cell types, the sieve elements, the parenchyma cells, and the sclerenchyma cells (**Figure 2a–d**). Sclerenchyma cells might sometimes be absent in primary and/or secondary phloem. The presence, quantities, and arrangements of these cell types in the tissue commonly vary and may be taxonomic informative [3, 4]. Lists depicting these variations in all phloem cell types are of ultimate importance for complete bark descriptions [5]. What follows is a description of these three major cell types in the phloem.

2.1 Conducting phloem cells

Sieve element is a general term that encompasses all conducting cells of the phloem, both sieve cells and sieve tube elements [1, 6]. The name sieve derives from the strainer appearance given to the cells by the presence of numerous pores crossing their bodies (**Figure 2c**). These pores are specialized plasmodesmata of wider diameter, and the sieve areas are basically specialized primary pit fields [7]. The sieve pores are usually lined up with callose, which were shown to be related with the formation of the sieve pores in angiosperms, although not in gymnosperms [8]. Large amounts of callose deposit in the sieve areas also when the sieve element loses conductivity, suffers injury, or becomes dormant. Callose in gymnosperms is typically wound callose [8]. Callose can be easily detected with aniline blue under fluorescence or resorcin blue [9] (**Figure 2b** and **c**).

Sieve elements have only primary walls, but sometimes this wall can be very thick receiving the name of nacreous walls (**Figure 2d**) [10] and can be present in all major vascular plant lineages [1]. Nacreous walls can be very thick, and some authors have proposed they would be secondary walls [1, 8]. Nacreous walls can almost occlude the entire lumen of the sieve element (**Figure 2d**); hence, its presence needs to be considered in experiments of sugar translocation. Such thick walls might be related to resistance to high turgor pressures within the sieve elements. Nacreous walls seem to have a strong phylogenetic signal and are much more common in some families, such as *Annonaceae*, *Calycanthaceae*, and *Magnoliaceae* [10].

There are basically two types of sieve elements: sieve cells and sieve tube elements. The sieve tube elements are distinguished by the presence of sieve plates, that is, sieve areas with wider and more abundant sieve pores, usually in both extreme ends of the cells, while sieve cells lack sieve plates [1, 6, 8]. A group of connected sieve tube elements form a sieve tube [8]. According to this concept, lycophytes and ferns have sieve cells [1]. However, because of the many differences in the morphology and distribution of protoplasm organelles and chemical substances between the sieve elements of gymnosperms and vascular cryptogams, Evert [8] suggests the use of "sieve cell" as exclusive to the gymnosperms, leaving the more general term "sieve element" to the lycophytes and ferns.

The longevity of sieve elements varies. In many species it is functional for just one growth season, while for other species they can be functional a couple of years, or in the case of plants that lack secondary growth, they will be living for the entire

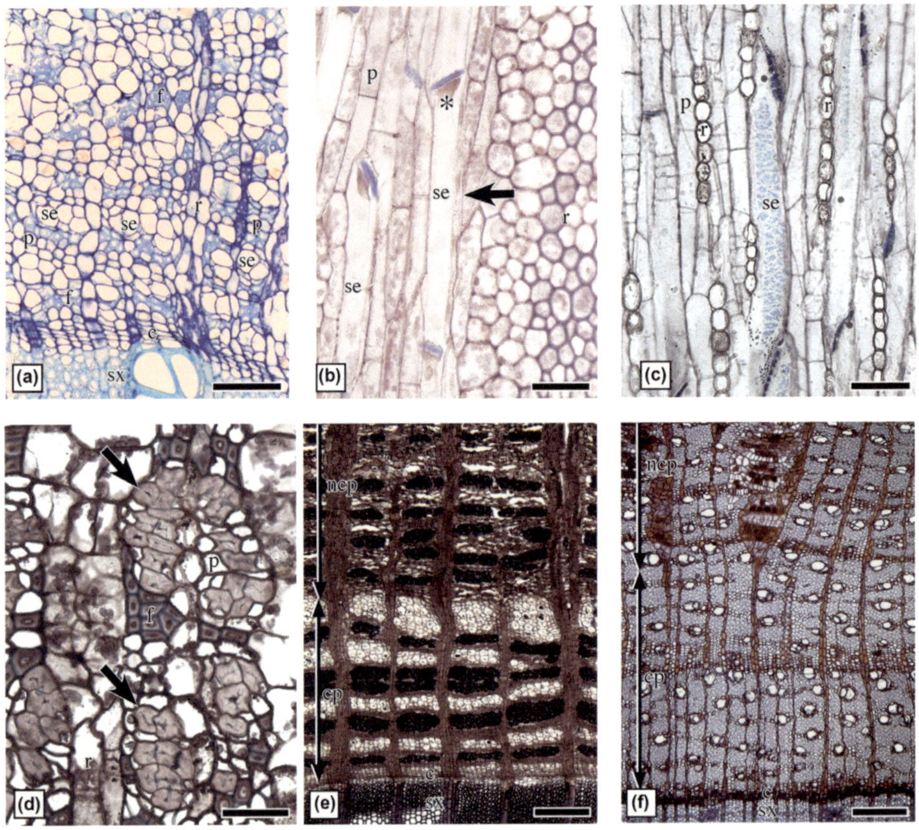

Figure 2.
General aspects of the secondary phloem. (a) Composition of the secondary phloem of Luehea divaricata (Malvaceae) TS, showing sieve tube elements (se) in clusters, axial parenchyma cells (p), fiber clusters (f), and rays (r). (b) Longitudinal tangential section (LT) of Cordia caffra (Boraginaceae) showing sieve tube element (se), companion cells (arrow), multiseriate ray (r), and axial parenchyma (p). Note callose staining with resorcin blue evidencing the slightly inclined simple sieve plates. Note also the P-protein (asterisk) next to the sieve plate. (c) LT of the secondary phloem of Castanea dentata (Fagaceae) showing sieve tube elements (se) with inclined, compound sieve plates and numerous lateral sieve areas of narrower pores, unicellular rays (r), and axial parenchyma (p). (d) TS of Talauma sp. (Magnoliaceae) showing sieve tube elements in clusters, with conspicuous nacreous walls, parenchyma cells (p), clusters of fibers (f), and rays (r). (e) Secondary phloem of maple, Acer saccharum (Sapindaceae), showing the conducting phloem (cp), where sieve tubes and companion cells are turgid, and the nonconducting phloem (ncp), with collapsed sieve tubes. (f) Secondary phloem of Carya cordiformis (Juglandaceae) showing a phloem formed by a background of fibers where solitary to multiple of two sieve tubes are scattered, with sieve-tube-centric and diffuse-in-aggregate axial parenchyma. Note that no collapse is seen in the nonconducting phloem of Carya. c, cambium; sx, secondary xylem. Scalebars: a = 100 μm, b-d = 50 μm, e, f = 200 μm.

plant life spam. Palm trees would perhaps be the plants with the oldest conducting sieve tube elements, since some reach 200 years [11]. In other plants, on the other hand, the sieve elements collapse a few cells away from the vascular cambium, corresponding to a fraction of the mm. In a mature tree, most of the secondary phloem will generally be composed of sieve elements no longer conducting. This region is called nonconducting phloem, in opposition to the area where sieve elements are turgid and conducting, called conducting phloem [5, 8] (**Figure 2e** and **f**). The term collapsed and noncollapsed phloem and functional and nonfunctional phloem are not recommended, since in some plants the nonconducting phloem keeps its sieve elements intact (**Figure 2f**), and although large parts of the phloem may not be conducting, the tissue as a whole is certainly still functioning in storage, protection, and even dividing or giving rise to new meristems, such as the phellogen and the dilatation meristem of some rays [5, 8].

2.1.1 Sieve cells and Strasburger cells

Sieve cells are typically very elongated cells with tapering ends (**Figure 3b**), which lack sieve plates, that is, lack an area in the sieve element where the pores are of a wider diameter. Even though the sieve areas may be more abundant in the terminal parts of the sieve cells, the pores in these terminal areas are of the same diameter as those of the lateral areas of the sieve element. Sieve cells lack P-protein in all stages of development. The sustenance of the sieve cells is carried by specialized parenchyma cells in close contact with the sieve elements, with numerous plasmodesmata, which maintain the physiological functioning of the sieve cells, including the loading and unloading of photosynthates. These cells are known either as albuminous cells or Strasburger cells. The name albuminous was initially coined given the proteinaceous appearance of these cell's contents. However, because the high protein content is not always present, the name Strasburger cell, paying tribute to its discoverer Erns Strasburger, is recommended over albuminous cells [5, 12]. Strasburger cells in the secondary phloem can be either axial parenchyma cells, as is common in *Ephedra* [13], or ray parenchyma cells, as is common in the conifers (**Figure 3c**) [14]. More commonly, the most conspicuous Strasburger cells in conifers are the marginal ray cells which are elongated (**Figure 3c**) and have a larger number of symplastic contact with the sieve cells [14]. Sometimes declining axial parenchyma cells also acts as Strasburger cells in *Pinus* [14]. The only reliable character to distinguish a Strasburger cell from an ordinary cell is the presence of conspicuous connections [14]. In the primary phloem, parenchyma cells next to the sieve cells are those which act as Strasburger cells.

2.1.2 Sieve tube elements and companion cells

A synapomorphy of the angiosperms is the presence of sieve tube elements and companion cells, both sister cells derived from the asymmetrical division of a single mother cell. In some instances, these mother cells can divide many times, creating assemblages of sieve tube elements and parenchyma cells

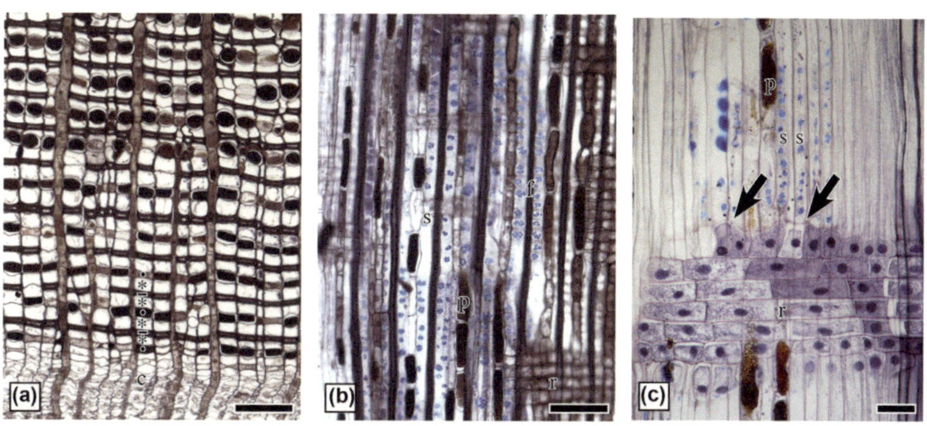

Figure 3.
*The secondary phloem of conifers. (a) Transverse section of the secondary phloem of Sequoia sempervirens (Cupressaceae) showing alternating tangential bands of sieve cells, axial parenchyma, and fibers, interrupted by uniseriate rays. (b). Longitudinal radial section (LR) of the secondary phloem of Sequoia sempervirens (Cupressaceae) showing alternating tangential bands of sieve cells, axial parenchyma, and fibers, interrupted by uniseriate rays. Sieve pores distributed across the walls of long sieve cells. (c). LR section of Pinus strobus (Pinaceae) showing the elongated marginal ray cells in close contact with the sieve cells. These are the Strasburger cells. f and rectangular symbol = fibers, s and * = the sieve cells, p and dot = axial parenchyma cells rich in tannins. Scalebars: a, b = 100 µm, c = 50 µm.*

ontogenetically related [15]. Sieve tube elements have specialized areas in the terminal parts of the sieve elements in which a sieve plate is present (**Figures 2b** and **c**). Within the sieve plate, the pores are much wider than those of the lateral sieve areas, evidencing a specialization of these areas for conduction [16]. In *Cucurbita*, the pores in the sieve plate have up to 10 μm in diameter, while the pores in the lateral sieve areas are of about 0.1 μm [7, 17]. The protoplast of sieve tube elements contain a specific constitutive protein called P-protein (P from phloem, also known as slime; **Figure 2b**), which in some taxa (e.g., *Leguminosae*) is nondispersive and can be seen as coagula inside of the sieve element [18].

Even in lineages of angiosperms where vessels were lost and tracheids re-evolved, such as *Winteraceae* in the *Magnoliids* and *Trochodendraceae* in the *eudicots*, sieve elements and companion cells are present [19], suggesting the independent evolution of these two plant vascular tissues derived from the same meristem initials. Since the sieve tube element loses its nucleus and ribosomes, the companion cell is the cell responsible for the metabolic life of the sieve elements, including the transport of carbohydrates in and out the sieve elements [7]. Companion cells may be arranged in vertical strands, with two to more cells (**Figure 2b**). Other parenchyma cells around the sieve tube integrate with the companion cells and can also act in this matter [7]. Typically, the cells closely related with the sieve tube elements die at the same time as the sieve element loses conductivity.

Sieve tube elements vary morphologically. The sieve plates can be transverse to slightly inclined (**Figure 2b**) or very inclined (**Figure 2c**) and contain a single sieve area (**Figure 2b**) or many (**Figure 2c**). When one sieve area is present, the sieve plate is named simple sieve plate, while when two to many are present, the sieve plates are called compound sieve plates. Compound sieve plates typically occur in sieve tube elements with inclined to very inclined sieve plates (**Figure 2c**). In addition, sieve elements with compound sieve plates are typically longer than those with simple sieve plates. Evolution to sieve elements of both sieve area types has been recorded in certain lineages, such as in *Arecaceae*, *Bignoniaceae*, and *Leguminosae* [5, 20], and to the present it is not still clear why the evolution of distinct morphologies would be or not beneficial. The only clear pattern is that compound sieve plates appear in long sieve elements [1], and phloem with a lot of fibers generally has compound sieve plates [20].

2.2 Parenchyma

In the primary phloem, just one type of parenchyma is present and typically intermingles with the sieve elements (**Figure 1d**). In the secondary structure, there are two types of parenchyma: axial parenchyma and ray parenchyma (**Figures 2b, c, 3b, c**), derived, respectively, from the fusiform and ray initials of the cambium.

The axial parenchyma in conifers commonly is arranged in concentric, alternating layers (**Figure 3a** and **b**). These parenchyma cells contain a lot of phenolic substances, which were viewed as a defense mechanism against bark attackers [21]. In Gnetales, the phloem axial parenchyma appears to be intermingling with the sieve cells (**Figure 4a**) [22]. Some of these axial parenchyma cells act as Strasburger cells [13].

In angiosperms, the distribution of the axial phloem parenchyma is more varied, and it may appear as a background tissue where other cells are dispersed or may be in bands (**Figure 4b** and **c**) and radial rows or sieve-tube-centric (**Figure 4d**) [5, 20]. The distribution of axial phloem parenchyma is commonly related to the abundance of fibers or sclereids. In species with more fibers, it is common to have a more organized arrangement of the parenchyma. For example, in *Robinia pseudoacacia* (*Leguminosae*) there are parenchyma bands in either side

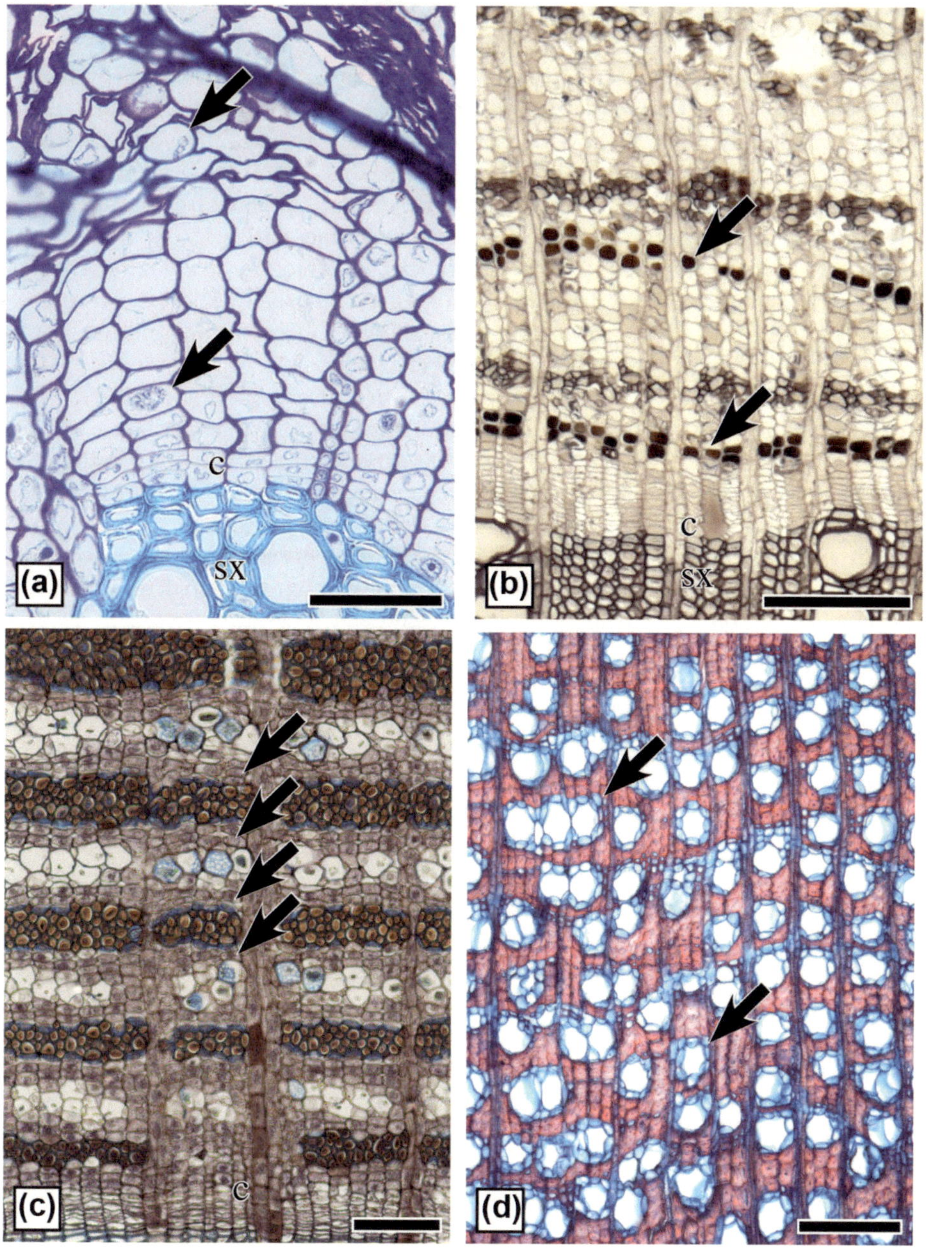

Figure 4.
Phloem axial parenchyma distribution in secondary phloem. (a) Ephedra tweediana (Ephedraceae) TS, showing sieve cells interspersed by axial parenchyma cells (arrows). Six to five cells away from the cambium, the sieve cells already lose conductivity and collapse with axial parenchyma cells enlarging (top arrow). (b) Lannea discolor (Anacardiaceae) TS showing axial parenchyma with tannins arranged in narrow bands (arrows). There are also other parenchyma cells with less content dispersed in the phloem. Note also the fibers in concentric bands. (c) Robinia pseudoacacia (Leguminosae) TS showing bands of axial parenchyma associated with the fiber bands and sieve tube elements in clusters with simple sieve plates staining with resorcin blue. (d) Fridericia nigrescens (Bignoniaceae) TS with sieve tubes surrounded by sieve-tube-centric axial parenchyma. The tissue background corresponds to the fibers. c, cambium; sx, secondary xylem; c, cambium; sp, secondary phloem; sx, secondary xylem. Scalebars: a = 50 µm; b, d = 200 µm; c = 100 µm.

of the concentric fiber bands (**Figure 4c**). When very large quantities of sclerenchyma are present, such as in the secondary phloem of *Carya* (*Juglandaceae*) or in *Fridericia*, *Tanaecium*, *Tynanthus*, and *Xylophragma* (*Bignoniaceae*), the sieve-tube-centric parenchyma appears (**Figure 4c**) and, as the name suggests, is surrounding the sieve tubes [8, 20, 23].

Although collectively described and referred to as axial phloem parenchyma, it is important to note that in many plants there will be distinct groups of phloem parenchyma within the phloem with quite different ergastic contents and therefore presumed different functions. Some of these specialized parenchyma cells may be considered secretory structures. Within a single plant, it is not uncommon that while some cells have crystals (especially when in contact with sclerenchyma), others have tannins, starch, and other substances. In apple trees (*Malus domestica*, *Rosaceae*) three types of axial parenchyma have been recorded: (1) crystal-bearing cells, (2) tannin- and starch-containing cells, and (3) those with no tannin or starch, which integrate with the companion cells [15].

Within bands of axial parenchyma, canals with a clear epithelium may be formed in many plant groups such as *Pinaceae*, *Anacardiaceae*, *Apiales*, a feature with strong phylogenetic signal. Some phloem parenchyma cells also act in the sustenance and support of the sieve elements, even when not derived from the same mother cell [7]. In longitudinal section, the axial phloem parenchyma may appear fusiform (not segmented) or in two up to several cells per strand [5].

While the phloem ages and moves away from the cambium, its structure dramatically change, and typically axial parenchyma cells enlarge (**Figures 4a** and **b**, **6c**), divide, and store more ergastic contents toward the nonconducting phloem. In plants with low fiber content, the dilatation undergone by the parenchyma cells typically provokes the collapse of the sieve elements. The axial parenchyma in the nonconducting phloem can dedifferentiate and give rise to new lateral meristems. In plants with multiple periderms, typically new phellogens are formed within the secondary phloem, compacting within the multiple periderms large quantities of dead, suberized phloem. In plants with variant secondary growth, especially lianas, new cambia might differentiate from axial phloem parenchyma cells [24]. In the Asian *Tetrastigma* (*Vitaceae*), new cambia were recorded differentiating from primary phloem parenchyma cells [25].

2.3 Sclerenchyma

Sclerenchymatic cells are those with thick secondary walls, commonly lignified. Sclerenchyma can be present or not in the phloem, and when present it typically gives structure to the tissue. For instance, a phloem with concentric layers of sclerenchyma cells is called stratified (**Figures 2e**, **3a**, and **4c**) [5]—not to be confused with storied, regarding the organization of the elements in tangential section. In Leguminosae, bands of phloem are associated to the concentric fiber bands (**Figure 4c**).

Older phloem shows more sclerification than younger phloem, and the sclerenchyma may also act as a barrier to bark attackers [21]. The sclerenchyma is typically divided in two categories: fibers and sclereids. These cell types differ mainly in form and size, but origin has also been used to distinguish them [26].

2.3.1 Fibers

Fibers are long and slender cells, derived from meristems, the fiber primordia [1, 26, 27]. In the primary phloem, fiber caps are sometimes found in association with the protophloem (**Figure 5a**) and are named protophloem fibers. Since only an ontogenetic study can evidence whether these fibers indeed differentiate within the protophloem, a term coined in the nineteenth century German and American literature, pericyclic fibers, has been recommended to be used instead of primary phloem fibers or perivascular fibers [5]. In the monocotyledons, fibers are commonly an important component of the vascular bundles (**Figure 5b–d**). Commonly these fibers are associated with the phloem (**Figure 5b**), but they might also be associated with the xylem (**Figure 5c**) or be central in the vascular bundle (**Figure 5d**). These fibers are not, however, understood as part of either phloem or xylem; although they are of vascular nature, they differentiate directly from procambium.

2.3.2 Sclereids

Sclereids may have different forms and sizes (**Figure 6a–c**). Within the phloem, they are more typically square or polygonal (stone cells) and contain numerous pits and conspicuous pit canals. Holdheid [26] defines that a sclereid is a cell derived from the belated sclerification of a parenchyma cell, and that is in fact the rule in the majority of cases (**Figure 6a** and **b**). However, there are lineages in which the sclereids differentiate very close to the cambium (e.g., *Pleonotoma*, *Bignoniaceae*, **Figure 6c**; [20]), and it would be untrue to claim that the derivatives had a stage as a mature parenchyma cell [1]. In these cases, the form is enough to define the sclereid.

On the other hand, there are cases where long and slender cells derive from previously mature parenchyma cells and are morphologically difficult to distinguish from fibers. In these cases, these cells are called fiber sclereids and may be even in concentric layers, such as in apple trees and pears (*Malus domestica* and *Pyrus*

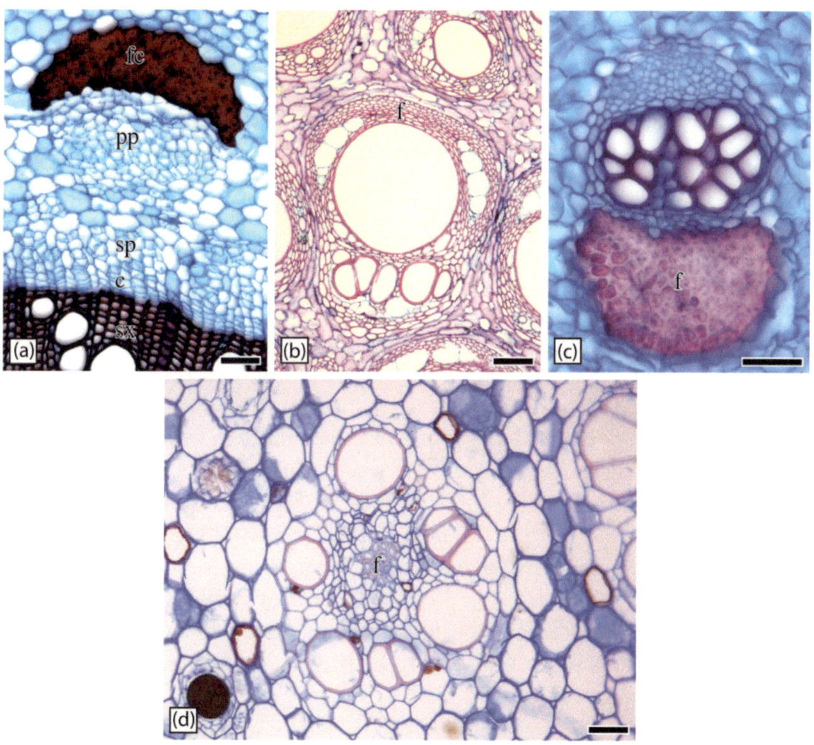

Figure 5.
Vascular fibers associated to eudicot and monocot primary structure. (a) Pericyclic fiber cap (fc) and primary phloem (pp) in Perianthomega vellozoi (Bignoniaceae). Secondary phloem (sp) beginning to be produced. Vascular bundles in monocotyledons. (b) Vascular bundle in the climber Calamus manan (Arecaceae) with fibers toward the phloem side. Phloem in two strands around a wide metaxylem vessel. (c) Vascular bundle of Vellozia alata (Velloziaceae), with fiber cap toward the xylem side. Phloem on the top side of the picture. (Picture credit to Marina Blanco Cattai). (d) Amphivasal bundle of Philodendron with fibers in the center of the vascular bundle and phloem surrounding it. Scalebars: a, b = 100 μm, c, d = 50 μm.

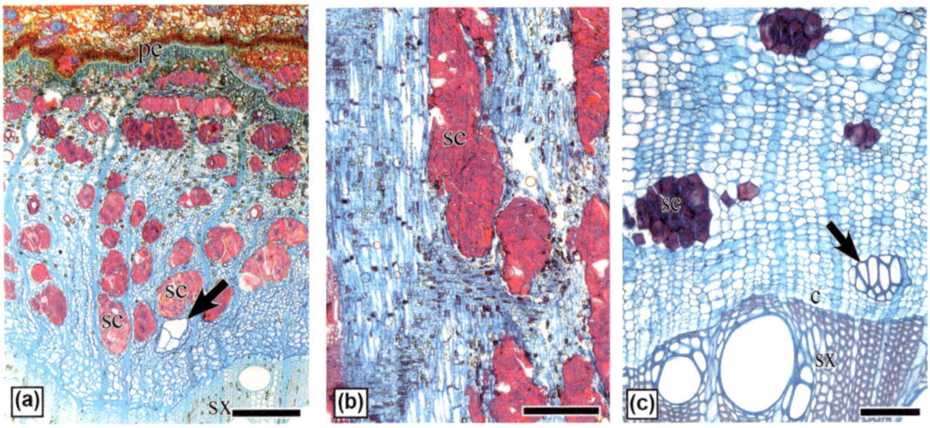

Figure 6.
Sclereids in the secondary phloem. (a) Sclereids (sc) differentiate from parenchyma cells (arrow) in the nonconducting phloem of Heteropterys intermedia (Malpighiaceae) TS, forming large clusters. (b) Longitudinal radial section of Heteropterys intermedia (Malpighiaceae) showing the sclereid masses. (c) In Pleonotoma tetraquetra (Bignoniaceae), the sclereids differentiate (arrow) close to the cambium within the conducting phloem. c, cambium; pe, periderm; sc, sclereid; sx, secondary xylem. Scalebars: a, b = 400 μm, c = 250 μm.

communis, respectively; [15]). Sclereids can also develop with different arrangements in the phloem, being isolated and scattered or in clusters (**Figures 6a–c**) [5].

2.4 Rays

The rays in the conducting phloem have typically the same organization in terms of width, height, and cellular composition as the secondary xylem. In this respect the rays vary from uniseriate to multiseriate (**Figure 7a**) and may be homocellular or heterocellular (**Figure 7b**). Homocellular rays are those composed of cells of one shape, all procumbent or all upright (common in many shrubs). Heterocellular rays are those where more than one cell shape is present together (**Figure 7b**). Ray composition is appreciated in radial sections.

Because the vascular cambium produces much more xylem to the inside than phloem to the outside, phloem rays typically greatly dilate toward the periphery of the organ (**Figure 7c**). It is not uncommon that a dilatation meristem longitudinal to the cambium forms in some barks (**Figure 7c**), especially in families with very wide, wedge-like rays such as the *Malvaceae*. Plants with unicellular rays very rarely have dilatation by cell division [15, 26]. Instead, they have great lateral expansion of their single cells. Ray width can be only determined in tangential sections.

Rays are typically exclusively parenchymatic; however, in many species sieve elements appear in the rays and are called ray sieve cells or radial sieve cells [5, 28, 29]. These cells were recorded connecting two different sieve tubes (collections of sieve tube elements). Ray sieve elements seem to be present in taxa where perforated ray cells have been also recorded [30].

3. Structure and development of primary and secondary phloem

3.1 Primary phloem

The primary phloem derives from the embryo in the seed and the procambium from the organ's apices. Similarly to the primary xylem, the primary phloem is

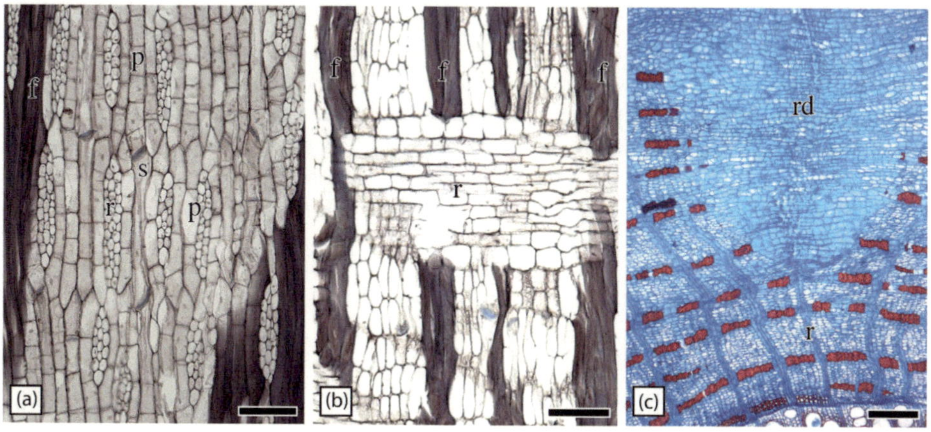

Figure 7.
Rays in the secondary phloem. (a) Longitudinal tangential section of Brachylaena transvaalensis (Asteraceae) showing storied structure, biseriate to triseriate rays (r), sieve tube elements with simple sieve plate (s) and axial parenchyma cells composed of 4–5 cells, and fibers. (b) Longitudinal radial section of Brachylaena transvaalensis (Asteraceae) showing heterocellular rays (r), with body procumbent and one row of marginal square cells. Fibers (f) in bands. (c) Ray dilatation (rd) by the formation of a dilatation meristem in the center of the ray in Perianthomega vellozoi (Bignoniaceae). f, fiber; p, axial parenchyma cell; r, ray; s, sieve tube element. Scalebars: a, b = 100 μm, c = 300 μm.

divided in protophloem and metaphloem (**Figure 1d**), with the protophloem differentiating first, while the plant is still elongating, and the metaphloem differentiating last. The phloem is always exarch, independently of the organ. Protophloem sieve elements sometimes lack companion cells, such as in *Arabidopsis*, and in this case the sieve elements are sustained by other neighboring parenchyma cells. Commonly, the protophloem quickly becomes obliterated and loses function. In plants without secondary growth, the metaphloem will be conducted during the entire life of the plant, as in the monocotyledons (**Figure 5b–d**) [11]. Different vascular plant lineages display different arrangements of the primary xylem and phloem, depending on the stele type. Two main types of steles exist, the protostele and the siphonostele. In the protostele, the entire center of the organ is composed of vascular tissue (**Figure 1a**), with the phloem in strands alternated with a central xylem in the protostele, haplostele, and actinostele (**Figure 1a**), while primary phloem is interspersed in the protostele plectostele [6]. The roots of all the vascular plants are protostelic (**Figure 1a**). The stems, however, can vary. In the lycophytes, they are always protostelic, while in the ferns (monilophytes) they might be protostelic, such as in *Psilotum*, or in all other range of siphonostelic steles [31]. The siphonostele evolved in concert with the macrophytes and resulted in the formation of a central pith derived from the ground meristem. No lineage displays as much diversity in the primary vasculature architecture as do the ferns. In the seed plants, that is, gymnosperms and angiosperms, the stem stele is always a syphonostele, either a eustele, where discrete vascular bundles form a concentric ring, or the atactostele, a type of stele exclusive of the monocotyledons where the bundles are scattered in the entire stem center. Some lineages of eudicotyledons and *Magnoliids* have evolved another subtype of siphonostele, the polycyclic eustele, where more than one ring of bundles is present, such as in *Piperaceae* and *Nyctaginaceae*.

The primary phloem is simpler than the secondary phloem and is basically formed by sieve elements and parenchyma cells (**Figure 1a–d**). Fiber caps are commonly present, and they might be phloematic (**Figure 5a**). For a discussion on their origin, check the section on fibers above. The position of the phloem is typically external or abaxial to the xylem, but in some lineages the bundles are bicollateral

(**Figure 1b**), and phloem is present both inside and outside (abaxial and adaxial), while in amphivasal bundles, the xylem encircles the phloem (**Figure 5d**), as in the secondary vascular tissues of some *Asparagales* [32, 33] and *Iridaceae* corms [34]. In some plant families and orders, intraxylary phloem (perimedullar phloem islands) is a synapomorphy, such as in the order *Myrtales* and in the families *Apocynaceae* and *Convolvulaceae* [35]. These phloem strands are initially primary, but a cambium can differentiate between the protoxylem and the phloem strands and develop secondary tissues inside of the pith.

3.2 Secondary phloem

Being derived from the cambium, the secondary phloem will share a number of characteristics with the secondary xylem. For instance, it is divided in an axial and radial system. The axial system is composed of sieve elements, axial parenchyma cells, and fibers, and the radial system is formed by rays, which are typically parenchymatic (**Figure 2a–c**). Similar to secondary xylem, the secondary phloem can be storied (**Figure 7a**) or non-storied (**Figure 2b** and **c**), depending whether the cambial mother cells are organized in tiers or not.

Some trees will have growth rings, with an early and a late phloem, both in temperate and tropical regions, but their characterization is only possible with periodical collections [5]. Sometimes, but not always, the fiber band width gives a hint on the presence of growth rings or the formation of very small sieve elements in the late phloem [1, 5].

3.2.1 Secondary phloem of gymnosperms

In conifers (except Gnetales) the secondary phloem is typically marked by an alternation of axial cell types (**Figure 3a** and **b**), uniseriate rays, and, in many lineages, axial and radial resin canals (e.g., *Pinaceae* and *Cupressaceae*). In the *Pinaceae*, the phloem is marked by the presence of an alternation of sieve cells and bands of axial parenchyma with phenolic contents, some also with druses. In the nonconducting phloem of *Pinaceae*, sclereids differentiate. In all other conifers, in addition to the alternation of parenchyma bands and sieve cells, fiber bands are present (**Figure 3a** and **b**). Therefore, sieve cells, parenchyma cells with phenolic content, and bands of fibers appear in alternation in non-*Pinaceae* and Gnetales conifers, including *Araucariaceae*, *Cupressaceae*, *Podocarpaceae*, *Taxaceae*, and *Taxodiaceae* [8, 21]. Another marked difference of these conifers compared to *Pinaceae* is that they contain a lot of crystals in their cell walls, including in Gnetales (see New World *Ephedra*; [36]), while in *Pinaceae* they are exclusively inside of idioblastic cells.

In other gymnosperms, in particular in Gnetales and Cycads, the first remarkable difference is the presence of very wide, multiseriate rays alternating with uniseriate rays. The wide rays in both groups have, however, evolved independently, since Cycads are a sister to all other gymnosperms, while Gnetales are within the conifers, as sister to the Pinaceae [31, 37]. In *Cyca* and the extinct *Cycadoidea*, sieve cells and phloem parenchyma alternate with fibers, which can be in tangential bands or not [38, 39]. In *Cyca*, the sieve cells appear in radial rolls [38], while in *Cycadoidea* there is a constant alternation of one sieve cell or phloem parenchyma to one fiber [39]. The nonconducting phloem of *Cycas* is marked by the collapse of sieve cells, enlargement of the axial parenchyma cells, ray dilatation, and sclerosis of some parenchymatic cells [38]. More than one ring of secondary phloem is present in some Cycads (e.g., *Cycas, Encephalartos*, *Lepidozamia*, and *Macrozamia*) and Gnetales (e.g., *Gnetum*), given that they have successive cambia [38, 40].

Within the Gnetales, in *Ephedra* axial parenchyma cells are interspersed with sieve cells (**Figure 4a**), and fiber may or may not be present and are typically gelatinous [36]. Fiber sclereids and/or sclereids appear in the nonconducting phloem of other species [13, 22]. In the nonconducting phloem of *Ephedra*, the sieve cells and Strasburger cells collapse with the enlargement of the axial and radial parenchyma cells (**Figure 4a**) with more ergastic contents [13]. In *Gnetum*, large areas of parenchyma sclerify, forming bands in the nonconducting phloem. The secondary phloem of *Welwitschia* is described as containing a large amount of fibers [21].

3.2.2 Secondary phloem of angiosperms

Within the angiosperms, the diversity of phloem cell type arrangements reaches its maximum. The structure can be storied (**Figure 7a**) or non-storied (**Figure 2b** and **c**); sclerenchyma can be present or lacking. The rays may be uni-, bi-, or multiseriate. A large array of secretory cells may be encountered, such as resin canals, laticifers, and mucilaginous cells. Crystalliferous parenchyma is also very common, especially when associated with fibers.

The variation in cell type arrangements can be of taxonomic interest. Sieve elements can vary in morphology and arrangement. They can be solitary (**Figure 2f**), scattered in the phloem (e.g., *Eucalyptus*, *Myrtaceae*), in clusters (e.g., *Malvaceae*; **Figures 2a**, **d** and **4c**), and in radial or tangential rows (many *Bignoniaceae*; [20]; **Figure 4d**). The functional significance of the different arrangements is unknown to date, although this is one of the features in the phloem with the strongest phylogenetic signal.

The presence, type, and arrangements of fibers and sclereids are one of the most informative characters in the bark [4]. In *Apocynaceae*, the fibers are completely absent, except in *Aspidosperma*, the sister group of all other *Apocynaceae* [35]. In *Aspidosperma*, they can appear solitary scattered across the phloem or in clusters. In some lineages, fibers appear in concentric alternating bands, as in *Leguminosae* (*Papilionoideae*), *Mimosoideae* (**Figure 4c**) [41], *Bignoniaceae* [20], and *Malvaceae*, and this is a constant character among them.

Phloem parenchyma more commonly constitute the background tissue in the phloem but can also be distributed in bands (**Figure 4b** and **c**), radial rows, or even only around the sieve tube elements (**Figure 4d**) [5].

4. Phloem activity

The classic theory of phloem transport is that proposed by Ernst Münch [42], and it involves the formation of an osmotic pressure transport gradient, where certain zones act as sources of sugars (leaves and storage organs), while others act as sinks. Experiments showed that the concentration gradients were always seen to be positive in the direction of flow [43], supporting Münch's postulate. In a system where transport goes against the direction of transpiration, its functionality relies on the presence of a plasma membrane across the entire system to create an osmotic pressure, hence the need of a conducting system with living cells [44]. Recent studies have been refining aspects involved in the photosynthate conduction to explain long-distance transports across large trees with such a simple system [44, 45]. A direct role of intracellular calcium has also been reported in the dissolution of nondispersive P-proteins and facilitation of transport [46]. Likely, the anatomical structure of the phloem discussed in the previous sections of this chapter will prove to play a role in the system. For instance, phloem sieve element length scale with the tree sizes and sieve plate type [45]. It was also shown that sieve element's diameter, length, and pore width increase from the top to the base of the trees [47, 48].

Across the entire pathway, sugars are removed from the system to sustain all cells in the plant body. This mechanism is only possible with the concerted mechanism between sieve elements and their close related cells (Strasburger cells and companion cells), with these accompanying cells constantly channeling substances and macromolecules toward the sieve elements [44]. The Strasburger and companion cells carry the loading and unloading of the sieve elements. Given the function of loading and unloading, the companion cell-sieve tube element size ratio is directly related to being in the source or the sink of sugars [44]. For instance, in leaves the companion cells are typically much larger, for they have the high demand of constantly loading the sieve tubes. In areas of release of the sugars (unloading), the companion cells are much smaller or even absent [44].

5. Economic uses

In the economic uses, it is not always easy to distinguish the use of the phloem from that of the periderm, since both together compose the bark of a woody plant. The phloem corresponds to the inner bark, and the periderm to the outer bark. The bark has a long history of utilization, from the production of remedies [49], aphrodisiacs (yohimbe), insecticides [50], dyes, tannins [50], angostura, fibers [51], gums and resins [50], latex, and flavorings [52].

In indigenous groups from British Columbia (Canada) and Tanzania, barks from dozens of species of woody plants are used as carbohydrate food, medicine, fibers, and structural material [50, 53]. In Mexico the bark of *Ficus* is used since prehispanic times to create a type of paper called *papel amate* (from the náhuatl paper = ámatl), used, for example, to create the Aztec codices.

The rubber tree, *Hevea brasiliensis* (*Euphorbiaceae*), is known from the extraction of latex to the production of rubber. Laticifers are present in concentric rings in the secondary phloem of the rubber tree and are an important economic asset in some tropical countries. Bark residues have also been considered for mulching [53–55], to build particle boards [56, 57], as fuel, and a source of food for ruminants [52].

Acknowledgements

I would like to express gratitude to Ray F. Evert, Veronica Angyalossy, Carmen Marcati, and André C. Lima for allowing their slide collections to be photographed and Leyde N. Nunes for the photo of *Tetrapterys* leaf, Solange Mazzoni Viveiros for photos of *Cucurbita*, and Marina Blanco Cattai for picture of *Vellozia*.

Conflict of interest

The authors declare no conflict of interest.

Author details

Marcelo R. Pace
Departamento de Botánica, Instituto de Biología, Universidad Nacional Autónoma de México, Mexico City, Mexico

*Address all correspondence to: marcelo.pace@ib.unam.mx

References

[1] Esau K. The Phloem. Encyclopedia of Plant Anatomy. Berlin: Gebrüder Borntraeger; 1969. 505 p

[2] Drobnitch ST, Jensen KH, Prentice P, Pittermann J. Convergent evolution of vascular optimization in kelp (*Laminariales*). Proceedings of the Royal Society B. 2015;**282**:20151667. DOI: 10.1098/rspb.2015.1667

[3] Zahur MS. Comparative Study of Secondary Phloem of 423 Species of Woody Dicotyledons Belonging to 85 Families. Ithaca: New York College of Agriculture at Cornell University; 1956. 160p

[4] Roth I. Structural Patterns of Tropical Barks. Berlin: Gebrüder Borntraeger; 1981. 609 p

[5] Angyalossy V, Pace MR, Evert RF, Marcati CR, Oskolski AA, Terrazas T, et al. IAWA list of microscopic bark features. IAWA Journal. 2016;**37**:517-615. DOI: 10.1163/22941932-20160151

[6] Eames AJ, MacDaniels LH. An Introduction to Plant Anatomy. 2nd ed. New York: McGraw-Hill; 1947. 427 p

[7] Esau K, Currier HB, Cheadle VI. Physiology of the phloem. Annual Review of Plant Physiology. 1957;**8**:349-374. DOI: 10.1146/annurev.pp.08.060157.002025

[8] Evert R. Esau's Plant Anatomy: Meristems, Cells, and Tissues of the Plant Body—Their Structure, Function, and Development. 3rd ed. Hoboken: Wiley; 2006. 601 p

[9] Cheadle V, Gifford EM, Esau K. A staining combination for phloem and contiguous tissues. Stain Technology. 1953;**28**:49-53. DOI: 10.3109/10520295309105101

[10] Esau K, Cheadle VI. Wall thickening in sieve elements. PNAS. 1958;**44**:546-553. DOI: 10.1073/pnas.44.6 546

[11] Tomlinson PB. The uniqueness of palms. Botanical Journal of the Linnean Society. 2006;**151**:5-14. DOI: 10.1111/j.1095-8339.2006.00520.x

[12] Trockenbrodt M. Survey and discussion of the terminology used in bark anatomy. IAWA Bulletin. 1990;**11**:141-166. DOI: 10.1163/22941932-90000511

[13] Alosi MC, Alfieri FJ. Ontogeny and structure of the secondary phloem in *Ephedra*. American Journal of Botany. 1972;**59**:818-827. DOI: 10.1002/j.1537-2197.1972.tb10156.x

[14] Alfieri FJ, Evert RF. Observations on albuminous cells in *Pinus*. Planta. 1968;**78**:93-97. DOI: 10.1007/BF00406643

[15] Evert RF. Ontogeny and structure of the secondary phloem of *Pyrus malus*. American Journal of Botany. 1963;**50**:8-37. DOI: 10.2307/2439857

[16] Esau K, Cheadle VI. Size of pores and their contents in sieve elements of dicotyledons. PNAS. 1959;**45**:156-162. DOI: 10.1073/pnas.45.2.156

[17] Volz G. Elektronenmikroskopische Untersuchungen über die Porengrössen pflanzlicher Zellwände. Mikroskopie. 1952;7:251-266

[18] Behnke HD. Distribution and evolution of forms and types of sieve-element plastids in the dicotyledons. Aliso. 1991;**13**:167-182. DOI: 10.5642/aliso.19911301.06

[19] Esau K, Cheadle VI. Anatomy of the secondary phloem in *Winteraceae*. IAWA Bulletin. 1984;**5**:13-42. DOI: 10.1163/22941932-90000853

[20] Pace MR, Alcantara S, Lohmann LG, Angyalossy V. Secondary phloem diversity and evolution in Bignonieae (*Bignoniaceae*). Annals of Botany. 2015;**116**:333-358. DOI: 10.1093/aob/mcv106

[21] Franceschi VR, Krokene P, Christiansen E, Krekling T. Anatomical and chemical defenses of conifer bark against bark beetles and other pests. New Phytologist. 2005;**167**:353-375

[22] Carlquist S. Wood, bark, and stem anatomy of gnetales: A summary. International Journal of Plant Sciences. 1996;**157**:S58-S76

[23] Jenke RU. Secondary Phloem Structure and Development in *Carya ovata*. La Crosse: Wisconsin State University; 1971

[24] Nair MNB, Mohan Ram HY. Structure of wood and cambial variant in the stem of *Dalbergia paniculate* Roxb. IAWA Bulletin. 1990;**11**:379-391

[25] Pace MR, Angyalossy V, Acevedo-Rodríguez P, Wen J. Structure and ontogeny of successive cambia in *Tetrastigma* (*Vitaceae*), the host plants of *Rafflesiaceae*. Journal of Systematics and Evolution. 2018;**56**:394-400. DOI: 10.1111/jse.12303

[26] Holdheid W. Anatomie mitteleuropäischer Geholzrinden. In: Freund H, editor. Handbuch der Mikroskopie in der Technik. Vol. 5. Frankfurt am Main: Umschau Verlag; 1951. pp. 193-367

[27] Esau K. Phloem. In: Metcalfe CR, Chalk L, editors. Anatomy of Dicotyledons. Systematic Anatomy of Leaf and Stem with a Brief History of the Subject. 2nd. ed. Oxford: Clarendon Press; 1979. pp. 181-189

[28] Rajput KS, Rao KS. Occurrence of sieve elements in phloem rays. IAWA Journal. 1997;**18**:197-201. DOI: 10.1163/22941932-90001479

[29] Rajput KS. Occurrence of radial sieve elements in the secondary phloem rays of some tropical species. Israel Journal of Plant Sciences. 2004;**52**:109-114. DOI: 10.1560/906M-ULMY-P5WK-9RAX

[30] Angyalossy V, Angeles G, Pace MR, Lima AC, Dias-Leme CL, Dias-Leme CL, et al. An overview of the anatomy, development and evolution of the vascular system of lianas. Plant Ecology and Diversity. 2012;**5**:167-182. DOI: 10.1080/17550874.2011.615574

[31] Simpson MG. Plant Systematics. Burlington: Elsevier Academic Press; 2006. p. 590

[32] Diggle PK, DeMason DA. The relationship between the primary thickening meristem and the secondary thickening meristem in *Yucca whipplei* Torr. I. Histology of the mature vegetative stem. American Journal of Botany. 1983;**70**:1195-1204. DOI: 10.2307/2443289

[33] Cattai MB, Menezes NL. Primary and secondary thickening in the stem of *Cordyline fruticosa* (*Agavaceae*). Anais da Academia Brasileira de Ciências. 2010;**82**:653-662. DOI: 10.1590/S0001-37652010000300013

[34] Chueiri-Chiaretto IA. Estrutura secundária do cormo de *Trimezia* Salisb. Ex herb. (*Iridaceae*). Ciência e Cultura. 1987;**39**:651-654

[35] Metcalfe CR, Chalk L. Anatomy of the Dicotyledons: Leaves, Stems, and Wood in Relation to Taxonomy with Notes on Economic Uses. Oxford: Clarendon Press; 1950

[36] Carlquist S. Wood and bark anatomy of the New World species of *Ephedra*. Aliso. 1988;**12**:441-483. DOI: 10.5642/aliso.19891203.04

[37] Lu Y, Ran JH, Guo DM, Yang ZY, Wang XQ. Phylogeny and divergence times of gymnosperms inferred from single-copy nuclear genes. PLoS One. 2014;**9**:e107679. DOI: 10.1371/journal.pone.0107679

[38] Terrazas T. Origin and activity of successive cambia in *Cycas* (*Cycadales*). American Journal of Botany. 1991;**78**:1335-1344. DOI: 10.2307/2445272

[39] Ryberg PE, Taylor EL, Taylor TN. Secondary phloem anatomy of *Cycadeoidea* (*Bennettitales*). American Journal of Botany. 2007;**94**:791-798. DOI: 10.3732/ajb.94.5.791

[40] Rao LN. Polyxyly in *Cycas circinalis* and *Cycas beddomei*, dyer. Current Science. 1972;**41**:793-797

[41] Costa CG, Coradin VTR, Czarneski CM, Pereira BAD. Bark anatomy of arborescent *Leguminosae* of Cerrado and gallery forest of Central Brazil. IAWA Journal. 1997;**18**:385-399. DOI: 10.1163/22941932-90001504

[42] Münch E. Material Flow in Plants. Gustav Fischer: Jena; 1930

[43] Zimmermann MH. Movement of organic substances in trees. Science. 1961;**133**:73-79. DOI: 10.1126/science.133.3446.73

[44] Van Bel AJE. The phloem, a miracle of ingenuity. Plant, Cell & Environment. 2003;**26**:125-149. DOI: 10.1046/j.1365-3040.2003.00963.x

[45] Liesche J, Pace MR, Xu Q, Li Y, Chen S. Height-related scaling of phloem anatomy and the evolution of sieve element end wall types in woody plants. New Phytologist. 2017;**214**:245-256. DOI: 10.1111/nph.14360

[46] Knoblauch M, Peters WS, Ehlers K, van Bel JE. Reversible calcium-regulated stopcocks in legume sieve tubes. The Plant Cell. 2001;**13**:1221-1230. DOI: 10.2307/3871375

[47] Liesche J, Schulz A. Phloem transport in gymnosperms: A question of pressure and resistance. Current Opinion in Plant Biology. 2018;**43**:36-42. DOI: 10.1016/j.pbi.2017.12.006

[48] Savage JA, Beecher SD, Clerx L, Gersony JT, Knoblauch J, Losada JM, et al. Maintenance of carbohydrate transport in tall trees. Nature Plants. 2017;**3**:965-972. DOI: 10.1038/s41477-017-0064-y

[49] Rihn B, Saliou C, Bottin MC, Keith G, Packer L. From ancient remedies to modern therapeutics: Pine bark uses in skin disorders revisited. Phytotherapy Research. 2001;**15**:76-78. DOI: /10.1002/1099-1573(200102)15:1<76::AID-PTR747>3.0.CO;2-O

[50] Gangol DR. Economic uses of forest plant resources in western Chitwan, Nepal. Banko Janakari. 2002;**12**:56-64

[51] Bender F. Spruce and balsam bark as a source of fiber products. Pulp and Paper Magazine of Canada. 1959;**60**:T275-T278

[52] Harkin JM, Rowe JW. Bark and its Possible Uses. Madison: FPL US Department of Agriculture Note 091. p. 56

[53] Gottesfeld LMJ. The importance of bark products in the aboriginal economies of Northwestern British Columbia, Canada. Economic Botany. 1992;**46**:148-157. DOI: 10.1007/BF02930629

[54] Bollen WB, Glennie DW. Fortified bark for mulching and soil conditioning. Forest Products Journal. 1963;**13**:209-215

[55] Basham BM, Thompson WS. An Economic Study of the Production and Use of Sawdust and Bark as Mulches and Soil Amendments for Horticultural

and Agricultural Purposes. Infor. Ser. No. 6, State College: Mississippi Forest Products Lab; p. 25

[56] Burrows CH. Particleboard from Douglas-fir Bark Without Additives. Information Circular 15. Corvallis: Oregon State University Forest Research Laboratory; p. 40

[57] Anderson AB, Breuer RJ, Nicholls GA. Bonding particle boards with bark extracts. Forest Products Journal. 1961;**11**:226-227

Chapter 2

Phytohormone-Mediated Homeostasis of Root System Architecture

Dongyang Xu and Masaaki K. Watahiki

Abstract

Unlike animals, most of the plants are sessile. This may be a reason why they developed the powerful ability of organ generation throughout their lifetime, which is distinct from the animals, whose generation potential is restricted in a certain period during development. Half part of the plant body, the root system, is hidden under the ground, where there is a competition of resources, for example, water and nutrients or biotic stresses and abiotic stresses surrounding the root system. With its strong regeneration ability, the architecture of the root system is shaped by all of these environmental cues together with the internal developmental signals. In this process, phytohormones work as the regulatory molecules mediating the internal and external developmental signals, thus controlling the morphology and function of the root system architecture. This chapter introduces the development of root system regulated by various phytohormones, like auxin, cytokinin, etc.

Keywords: root architecture, auxin, cytokinin, postembryonic organogenesis

1. Anatomy and development of root

1.1 Root system architecture

In different plant species, root system architecture (RSA) has diverse morphologies. There are basically two types of RSA, the taproot system (or allorhizic system) in gymnosperms and dicotyledons, like *Arabidopsis thaliana* (*Arabidopsis*), tomato (*Solanum lycopersicum* L.), carrot (*Daucus carota*), and poplar (*Populus* spp.) and the fibrous root system (or homorhizic system) in monocotyledons such as maize (*Zea mays* L.), rice (*Oryza sativa* L.), onion (*Allium cepa*), garlic (*Allium sativum*), and tulip (*Tulipa* spp.) [1]. The taproot system consists of a single thick central primary root (PR) with thin or no lateral roots (LRs); the fibrous root system has small and short-lived primary and adventitious roots (ARs) derived from shoots, stems, or leaves [1].

1.2 Intrinsic developmental signals and environmental conditions modify root system architecture

Arabidopsis as a eudicot has a taproot system, which consists of an embryonic radicle-derived PR and postembryonic-developed LRs and ARs. Root regeneration exists throughout the plants' lifetime; it is a distinctive feature of plants and contributes to their robustness in adverse conditions.

In *Arabidopsis*, LRs initiate from pairs of pericycle cells that possess developmental potential as plant stem cells. These pericycle cells are selected and directed to become LR founder cells and form LRs by both intrinsic and environmental signals [2–5]. The primary LR is initiated from the basal meristem of the PR, where root cap-derived auxin influences the amplitude of oscillatory gene expression in the basal meristem and the elongation zone of the root, which leads to the pre-patterning of LR initiation sites [6, 7]. The pre-patterning process is marked by the expression of a series of genes, like *GATA23*, *MEMBRANE-ASSOCIATED KINASE REGULATOR 4* (*MAKR4*), and *IAA19* [7]. In the basal meristem and elongation zone, *DR5::Luciferase* expression was observed to rhythmically pulse with a period of ~6 h, which matched with the period of LR pre-branch site production [6]. It is recently reported that the source of auxin is provided by the cyclic programmed cell death of root cap cells [8, 9].

It is noteworthy that not all of the pre-branch sites emerge to be LRs [6]. The dormant pre-branch sites may present a selective mechanism for LR formation under certain growth conditions, such as water availability, nutrient levels, physical obstacles, or damage [5, 10–13]. It is interesting that many of the external signals converge on phytohormones to regulate root development. Among these phytohormones, auxin functions as a central mediator.

Mechanical forces are important regulators for plant morphogenesis. LRs always emerge from the convex side of PR bending, resulting in a left-right alternation of LRs. Bending caused by gravitropic curvature led to the initiation of LRs, where a subcellular relocalization of PIN1 was observed [11]. Release the pericycle cells from the restraints of adjacent endodermis by targeted single cell ablation of endodermal cells triggered the pericycle to reenter the cell cycle and induced auxin-dependent LR initiation [14]. Excision of the *Arabidopsis* PR leads to the promotion of LR formation, which is mediated by activated auxin biosynthesis and auxin transport [15].

2. Roles of phytohormones on root formation

2.1 Auxin

The phytohormone auxin which plays fundamental roles in many aspects of plant growth and development is also a well-documented key regulator of LR development [16, 17]. The natural auxin, indole-3-acetic acid (IAA), is mainly synthesized in a two-step pathway from tryptophan. First, tryptophan is converted to indole-3-pyruvate (IPA) by the TAA1/SAV3 family of aminotransferases; IPA is then converted to IAA by the YUCCA (YUC) family of flavin monooxygenases [18–23]. Auxin biosynthesis has been shown to play an essential role on both programed and wound-induced LR and AR developments [15, 24, 25].

Polar auxin transport (PAT), mediated by auxin influx (AUX1 and LAXs) and efflux carriers (PINs and MDR/PGPs) [26–29], generates auxin gradients and maintains an auxin maximum to regulate LR formation and positioning [17, 30–33].

Auxin signaling is known to be an integrator of endogenous and exogenous signals for root branching [17, 30, 34, 35]. It begins with the degradation of a class of AUXIN/INDOLE-3-ACETIC ACID (Aux/IAA) through TRANSPORT INHIBITOR RESPONSE 1 (TIR1) auxin receptor [36, 37], resulting in the activation of the AUXIN RESPONSE FACTOR (ARF) [38, 39]. ARF7 and ARF19 transcription factors further induce the expression of downstream target genes like *LATERAL*

ORGAN BOUNDARIES DOMAIN/ASYMMETRIC LEAVES2-LIKE (*LBD/ASL*) family genes *LBD16/ASL18* and *LBD29/ASL16*, promoting LR initiation at the protoxylem-pole pericycle cells [40–43].

2.2 Cytokinin

Cytokinin is also a main player in root development. In higher plants, isopentenyladenine (iP), trans-zeatin (tZ), and dihydrozeatin (dZ) are the predominant cytokinins [44]. Cytokinin level and patterning in plant are controlled by a fine equilibrium between cytokinin synthesis and catabolism [44, 45]. Cytokinin biosynthesis is dependent on the activity of *ATP/ADP-isopentenyltransferase* (*IPT*) and *LONELY GUY* (*LOG*) gene family [46–48], and the metabolism is mainly through the *CYTOKININ OXIDASE/DEHYDROGENASE* (*CKX*) genes [44, 45]. Cytokinin can also be inactivated through conjugation to glucose [49]. The spatial and temporal distribution of cytokinin is in part due to the specific expression of cytokinin synthesis and catabolism genes [45, 47, 48, 50–52].

In *Arabidopsis*, cytokinin signaling starts with the perception by the transmembrane cytokinin receptors ARABIDOPSIS HISTIDINE KINASE (AHK), AHK2, AHK3, and AHK4/WOL1/CRE1 [53–55], which target the histidine phosphotransfer protein AHPs to activate the type-A and type-B ARABIDOPSIS RESPONSE REGULATORS (ARRs) that negatively and positively regulate cytokinin signaling, respectively [55–61]. Two type-A ARRs, ARR7, and ARR15 were induced by both cytokine and auxin and are essential for embryonic root development [62].

Although some evidences showed that cytokinins act as both local and long-distance signals [51, 63–65], and some transporter proteins have been shown to be involved in cytokine transport [66–70], the molecular mechanisms of cytokinin transport are still not well characterized.

Postembryonic root development is regulated by the root apical meristem (RAM), where cytokinin is known to act antagonistically with auxin to control the balance of cell division in the division zone and cell differentiation in the transition zone, which is essential for the maintenance of the RAM and affects the growth and patterning of the root [64, 71]. Application of cytokinin reduces the number of meristem cells and the size of RAM and promotes cell differentiation in the transition zone; cytokinin biosynthesis and signaling mutants as well as *CKX* overexpression mutants have a larger RAM with more meristem cells [45, 64, 72]. Conversely, auxin treatment increases meristem size and promotes cell division in the proximal meristem, and auxin transporter *PIN* mutants display a smaller RAM [64, 73]. The cross-talk of cytokine and auxin in regulating RAM activity was shown to converge on the auxin-inducible *AUX/IAA* family gene *SHORT HYPOCOTYL 2/ INDOLE-3-ACETIC ACID 3* (*SHY2/IAA3*) in the transition zone, where cytokinin activates *SHY2* via the *AHK3/ARR1* two-component signaling pathway to suppress *PIN3* and *PIN7* expression and promote cell differentiation, while auxin suppressed SHY2 protein, leading to the activation of PINs and promotion of cell division [71]. Furthermore, Moubayidin et al. [74] revealed that in transition zone, SCR, a member of the GRAS family of transcription factors, directly represses the expression of *ARR1*, which controls auxin production via the *ASB1* gene and sustains stem cell activity, to simultaneously control stem cell division and differentiation and ensure coherent root growth. Cytokinin affects the expression of multiple *PINs* differentially in a tissue-specific manner to regulate auxin distribution [75, 76]. Auxin-cytokinin interactions lead the generation of distinct hormonal response zones, thus controlling the development of root vascular tissue.

On contrary to auxin, which is a positive regulator of LR development, cytokinin acts as a negative regulator of LR formation. Cytokinin suppresses LR initiation through downregulating *PIN* expression and preventing the establishment of auxin gradient in LR founder cells [77]. Mutants with reduced cytokinin level or deficient cytokinin signaling increased the number of LRs [45, 58, 60, 78], while cytokinin treatment suppresses LR initiation and development [77, 79, 80]. Li et al. [80] reported that cytokinins inhibit LR initiation by blocking the cell cycle of pericycle founder cell at G_2 to M transition phase while promoting LR elongation by stimulation of the G_1 to S transition.

Through mutant analysis Chang et al. [81] showed that cytokinin biosynthesis genes *IPT3* and *IPT5* and all three cytokinin receptor genes *AHK2*, *AHK3*, and *CRE1/AHK4* act redundantly during LR initiation, and early stages of lateral root primordia (LRP) formation are particularly cytokinin sensitive. They suggest that cytokinin may serve as a positional signal for new LRP formation. In rice, ERF3 interacts with WOX11 to promote crown root initiation and elongation by regulating the cytokinin-responsive gene *RR2* [82]. Cytokinin has also been shown to modulate LR formation by mediating environmental cues. Jeon et al. [83] showed that CYTOKININ RESPONSE FACTOR 2 (CRF2) and CRF3 encoding APETALA2 transcription factors regulate *Arabidopsis* LR initiation under cold stress.

2.3 Other phytohormones

Other phytohormones, like abscisic acid (ABA), gibberellic acid (GA), brassinosteroid (BR), jasmonic acid (JA), ethylene, and strigolactone (SL), also participate in root growth and development.

Signora et al. [84] showed that ABA plays an important role in mediating the effects of nitrate on LR formation in *Arabidopsis*. Brady et al. [85] reported that *ABSCISIC ACID INSENSITIVE 3* (*ABI3*) is involved in auxin signaling and LR development. De Smet et al. [86] reported that ABA application leads to the inhibition of LR development immediately after the emergence of the LRP from the parent root and prior to the activation of the LR meristem in an auxin-independent manner. Shkolnik-Inbar and Bar-Zvi [87] showed that *ABI4*, which encodes an ABA-regulated AP2 domain transcription factor, mediates ABA and cytokinin inhibition of LR formation through the reduction of PAT. Ding et al. [88] reported that ABA signaling and auxin homeostasis regulate WRKY46 to modulate the development of LR in osmotic/salt stress condition.

Hansen [89] reported on the GA-mediated light dependent promotion and inhibition of AR formation. Through mutant analysis, Yaxley et al. [90] showed that GA is important for normal root elongation in pea. Fu and Harberd [91] showed that auxin regulates root growth through GA-mediated DELLA protein destabilization. Steffens et al. [92] showed that GA is ineffective on its own but acts synergistically with ethylene to promote the number of penetrating roots and the growth rate of emerged roots in deepwater rice.

BR is a positive regulator of root development [93]. Bao et al. [94] showed that BRs interact with auxin to promote LR development.

JA, a crucial plant defense hormone, also participated in the regulation of root development. Raya-González et al. [95] observed that low concentrations of JA inhibited PR growth through an auxin-independent manner and promoted LR formation auxin-dependently, and JA receptor COI1 is involved in JA-induced LR formation and LR positioning. Cross-talk between JA and auxin has been frequently reported. JA has been reported to be implicated in YUC9-mediated auxin biosynthesis in wounded leaves in *Arabidopsis* [96]. Cai et al. [97] also reported a cross-talk between JA and auxin biosynthesis during LR formation mediated by ERF109. Gutierrez et al. [98] showed that auxin controls *Arabidopsis* AR initiation through the regulation of JA homeostasis.

Ethylene is also a well-known phytohormone that participates in the plant defense signaling pathways. Strader et al. [99] reported that ethylene interact with auxin to control root cell expansion. Ivanchenko et al. [100] observed application of low level of ethylene precursor 1-aminocyclopropane-1-carboxylic acid (ACC) promotes LRP initiation, while higher doses of ACC strongly inhibits LRP initiation but promotes LRP emergence; this regulation of LR initiation and emergence by ethylene is through interactions with auxin. Lewis et al. [101] reported that ethylene suppresses LR formation through promotion of PIN3 and PIN7-mediated auxin efflux to prevent local auxin accumulation.

Jiang et al. [102] showed that SL analog GR24 negatively influenced LR priming and emergence, which is dependent on the intimate connection with auxins and cytokinins, with the PAT capacity as a central player.

3. Conclusions

The root system of higher plants is modified by intrinsic developmental signals and diverse environmental cues. Both the internal and the external signals converged on phytohormones to regulate the formation of a highly plastic and adaptive RSA, which sustains the growth of plants even in adverse conditions. Several lines of evidences suggest that cross-talks among different phytohormones are essential for the regulation of root development, and auxin plays a central role in these processes. Although auxin and cytokinin as the key regulators of root development have been extensively studied, the roles of other phytohormones still need to be further characterized to give us a full view of plant root development.

Conflict of interest

The authors declare no conflict of interest.

Acronyms and abbreviations

RSA	root system architecture
PR	primary root
LR	lateral root
AR	adventitious root
IAA	indole-3-acetic acid
PAT	polar auxin transport
iP	isopentenyladenine
tZ	trans-zeatin
dZ	dihydrozeatin
RAM	root apical meristem
LRPs	lateral root primordia
ABA	abscisic acid
GA	gibberellic acid
BR	brassinosteroid
JA	jasmonic acid
SL	strigolactone
ACC	1-aminocyclopropane-1-carboxylic acid

Author details

Dongyang Xu[1] and Masaaki K. Watahiki[2*]

1 Institute of Genomics, Huaqiao University, Xiamen, China

2 Faculty of Science, Hokkaido University, Sapporo, Japan

*Address all correspondence to: watahiki@sci.hokudai.ac.jp

References

[1] Bellini C, Pacurar DI, Perrone I. Adventitious roots and lateral roots: Similarities and differences. Annual Review of Plant Biology. 2014;**65**:639-666

[2] De Smet I, Tetsumura T, De Rybel B, dit Frey NF, Laplaze L, Casimiro I, et al. Auxin-dependent regulation of lateral root positioning in the basal meristem of *Arabidopsis*. Development. 2007;**134**(4):681-690

[3] Dubrovsky JG, Sauer M, Napsucialy-Mendivil S, Ivanchenko MG, Friml J, Shishkova S, et al. Auxin acts as a local morphogenetic trigger to specify lateral root founder cells. Proceedings of the National Academy of Sciences. 2008;**105**(25):8790-8794

[4] Richter GL, Monshausen GB, Krol A, Gilroy S. Mechanical stimuli modulate lateral root organogenesis. Plant Physiology. 2009;**151**(4):1855-1866

[5] Sugimoto K, Jiao Y, Meyerowitz EM. *Arabidopsis* regeneration from multiple tissues occurs via a root development pathway. Developmental Cell. 2010;**18**(3):463-471

[6] Moreno-Risueno MA, Van Norman JM, Moreno A, Zhang J, Ahnert SE, Benfey PN. Oscillating gene expression determines competence for periodic *Arabidopsis* root branching. Science. 2010;**329**(5997):1306-1311

[7] Xuan W, Audenaert D, Parizot B, Möller BK, Njo MF, De Rybel B, et al. Root cap-derived auxin pre-patterns the longitudinal axis of the *Arabidopsis* root. Current Biology. 2015;**25**(10):1381-1388

[8] Xuan W, Band LR, Kumpf RP, Van Damme D, Parizot B, De Rop G, et al. Cyclic programmed cell death stimulates hormone signaling and root development in *Arabidopsis*. Science. 2016;**351**(6271):384-387

[9] Möller BK, Xuan W, Beeckman T. Dynamic control of lateral root positioning. Current Opinion in Plant Biology. 2017;**35**:1-7

[10] Al-Ghazi Y, Muller B, Pinloche S, Tranbarger T, Nacry P, Rossignol M, et al. Temporal responses of *Arabidopsis* root architecture to phosphate starvation: Evidence for the involvement of auxin signalling. Plant, Cell & Environment. 2003;**26**(7):1053-1066

[11] Ditengou FA, Teale WD, Kochersperger P, Flittner KA, Kneuper I, van der Graaff E, et al. Mechanical induction of lateral root initiation in *Arabidopsis thaliana*. Proceedings of the National Academy of Sciences. 2008;**105**(48):18818-18823. DOI: 10.1073/pnas.0807814105

[12] Sena G, Birnbaum KD. Built to rebuild: In search of organizing principles in plant regeneration. Current Opinion in Genetics & Development. 2010;**20**(4):460-465

[13] Van Norman JM, Xuan W, Beeckman T, Benfey PN. To branch or not to branch: The role of pre-patterning in lateral root formation. Development. 2013;**140**(21):4301-4310

[14] Marhavý P, Montesinos JC, Abuzeineh A, Van Damme D, Vermeer JE, Duclercq J, et al. Targeted cell elimination reveals an auxin-guided biphasic mode of lateral root initiation. Genes & Development. 2016;**30**(4):471-483

[15] Xu D, Miao J, Yumoto E, Yokota T, Asahina M, Watahiki M. YUCCA9-mediated auxin biosynthesis and polar auxin transport synergistically regulate regeneration of root systems following root cutting. Plant and Cell Physiology. 2017;**58**(10):1710-1723

[16] Fukaki H, Okushima Y, Tasaka M. Auxin-mediated lateral root formation

in higher plants. International Review of Cytology. 2007;**256**:111-137

[17] Lavenus J, Goh T, Roberts I, Guyomarc'h S, Lucas M, De Smet I, et al. Lateral root development in *Arabidopsis*: Fifty shades of auxin. Trends in Plant Science. 2013;**18**(8):450-458

[18] Tao Y, Ferrer J-L, Ljung K, Pojer F, Hong F, Long JA, et al. Rapid synthesis of auxin via a new tryptophan-dependent pathway is required for shade avoidance in plants. Cell. 2008;**133**(1):164-176

[19] Stepanova AN, Robertson-Hoyt J, Yun J, Benavente LM, Xie D-Y, Doležal K, et al. TAA1-mediated auxin biosynthesis is essential for hormone crosstalk and plant development. Cell. 2008;**133**(1):177-191

[20] Yamada M, Greenham K, Prigge MJ, Jensen PJ, Estelle M. The TRANSPORT INHIBITOR RESPONSE2 gene is required for auxin synthesis and diverse aspects of plant development. Plant Physiology. 2009;**151**(1):168-179

[21] Won C, Shen X, Mashiguchi K, Zheng Z, Dai X, Cheng Y, et al. Conversion of tryptophan to indole-3-acetic acid by TRYPTOPHAN AMINOTRANSFERASES OF *ARABIDOPSIS* and YUCCAs in *Arabidopsis*. Proceedings of the National Academy of Sciences. 2011;**108**(45):18518-18523

[22] Mashiguchi K, Tanaka K, Sakai T, Sugawara S, Kawaide H, Natsume M, et al. The main auxin biosynthesis pathway in *Arabidopsis*. Proceedings of the National Academy of Sciences. 2011;**108**(45):18512-18517

[23] Zhao Y. Auxin biosynthesis: A simple two-step pathway converts tryptophan to indole-3-acetic acid in plants. Molecular Plant. 2012;**5**(2):334-338

[24] Sheng L, Hu X, Du Y, Zhang G, Huang H, Scheres B, et al. Non-canonical WOX11-mediated root branching contributes to plasticity in *Arabidopsis* root system architecture. Development. 2017;**144**(17):3126-3133. DOI: 10.1242/dev.152132

[25] Chen Q , Dai X, De-Paoli H, Cheng Y, Takebayashi Y, Kasahara H, et al. Auxin overproduction in shoots cannot rescue auxin deficiencies in *Arabidopsis* roots. Plant & Cell Physiology. 2014;**55**(6):1072-1079

[26] Swarup R, Péret B. AUX/LAX family of auxin influx carriers—An overview. Frontiers in Plant Science. 2012;**3**:225

[27] Paponov IA, Teale WD, Trebar M, Blilou I, Palme K. The PIN auxin efflux facilitators: Evolutionary and functional perspectives. Trends in Plant Science. 2005;**10**(4):170-177

[28] Blakeslee JJ, Peer WA, Murphy AS. MDR/PGP auxin transport proteins and endocytic cycling. Plant Endocytosis. 2005;**1**:159-176

[29] Kramer EM, Bennett MJ. Auxin transport: A field in flux. Trends in Plant Science. 2006;**11**(8):382-386

[30] Casimiro I, Marchant A, Bhalerao RP, Beeckman T, Dhooge S, Swarup R, et al. Auxin transport promotes *Arabidopsis* lateral root initiation. The Plant Cell. 2001;**13**(4):843-852

[31] Marchant A, Bhalerao R, Casimiro I, Eklöf J, Casero PJ, Bennett M, et al. AUX1 promotes lateral root formation by facilitating indole-3-acetic acid distribution between sink and source tissues in the *Arabidopsis* seedling. The Plant Cell. 2002;**14**(3):589-597

[32] Benková E, Michniewicz M, Sauer M, Teichmann T, Seifertová D, Jürgens G, et al. Local, efflux-dependent auxin gradients as a common module

for plant organ formation. Cell. 2003;**115**(5):591-602

[33] De Smet I. Lateral root initiation: One step at a time. The New Phytologist. 2012;**193**(4):867-873

[34] De Smet I, Vanneste S, Inzé D, Beeckman T. Lateral root initiation or the birth of a new meristem. Plant Molecular Biology. 2006;**60**(6):871-887

[35] Fukaki H, Tasaka M. Hormone interactions during lateral root formation. Plant Molecular Biology. 2009;**69**(4):437

[36] Dharmasiri N, Dharmasiri S, Estelle M. The F-box protein TIR1 is an auxin receptor. Nature. 2005;**435**(7041):441

[37] Kepinski S, Leyser O. The *Arabidopsis* F-box protein TIR1 is an auxin receptor. Nature. 2005;**435**(7041):446

[38] Ulmasov T, Hagen G, Guilfoyle TJ. ARF1, a transcription factor that binds to auxin response elements. Science. 1997;**276**(5320):1865-1868

[39] Nanao MH, Vinos-Poyo T, Brunoud G, Thévenon E, Mazzoleni M, Mast D, et al. Structural basis for oligomerization of auxin transcriptional regulators. Nature Communications. 2014;**5**:3617

[40] Okushima Y, Overvoorde PJ, Arima K, Alonso JM, Chan A, Chang C, et al. Functional genomic analysis of the AUXIN RESPONSE FACTOR gene family members in *Arabidopsis thaliana*: Unique and overlapping functions of ARF7 and ARF19. The Plant Cell. 2005;**17**(2):444-463

[41] Okushima Y, Fukaki H, Onoda M, Theologis A, Tasaka M. ARF7 and ARF19 regulate lateral root formation via direct activation of LBD/ASL genes in *Arabidopsis*. The Plant Cell. 2007;**19**(1):118-130

[42] Goh T, Kasahara H, Mimura T, Kamiya Y, Fukaki H. Multiple AUX/IAA–ARF modules regulate lateral root formation: The role of *Arabidopsis* SHY2/IAA3-mediated auxin signalling. Philosophical Transactions of the Royal Society B. 2012;**367**(1595):1461-1468

[43] Lee HW, Kim NY, Lee DJ, Kim J. LBD18/ASL20 regulates lateral root formation in combination with LBD16/ASL18 downstream of ARF7 and ARF19 in *Arabidopsis*. Plant Physiology. 2009;**151**(3):1377-1389

[44] Sakakibara H. Cytokinins: Activity, biosynthesis, and translocation. Annual Review of Plant Biology. 2006;**57**:431-449

[45] Werner T, Motyka V, Laucou V, Smets R, Van Onckelen H, Schmülling T. Cytokinin-deficient transgenic *Arabidopsis* plants show multiple developmental alterations indicating opposite functions of cytokinins in the regulation of shoot and root meristem activity. The Plant Cell. 2003;**15**(11):2532-2550

[46] Kakimoto T. Identification of plant cytokinin biosynthetic enzymes as dimethylallyl diphosphate: ATP/ADP isopentenyltransferases. Plant & Cell Physiology. 2001;**42**(7):677-685

[47] Miyawaki K, Matsumoto-Kitano M, Kakimoto T. Expression of cytokinin biosynthetic isopentenyltransferase genes in *Arabidopsis*: Tissue specificity and regulation by auxin, cytokinin, and nitrate. The Plant Journal. 2004;**37**(1):128-138

[48] Kuroha T, Tokunaga H, Kojima M, Ueda N, Ishida T, Nagawa S, et al. Functional analyses of LONELY GUY cytokinin-activating enzymes reveal the importance of the direct activation pathway in *Arabidopsis*. The Plant Cell. 2009;**21**(10):3152-3169

[49] Bajguz A, Piotrowska A. Conjugates of auxin and cytokinin. Phytochemistry. 2009;**70**(8):957-969

[50] Miyawaki K, Tarkowski P, Matsumoto-Kitano M, Kato T, Sato S, Tarkowska D, et al. Roles of *Arabidopsis* ATP/ADP isopentenyltransferases and tRNA isopentenyltransferases in cytokinin biosynthesis. Proceedings of the National Academy of Sciences. 2006;**103**(44):16598-16603

[51] Kurakawa T, Ueda N, Maekawa M, Kobayashi K, Kojima M, Nagato Y, et al. Direct control of shoot meristem activity by a cytokinin-activating enzyme. Nature. 2007;**445**(7128):652

[52] Takei K, Sakakibara H, Sugiyama T. Identification of genes encoding adenylate isopentenyltransferase, a cytokinin biosynthesis enzyme, in *Arabidopsis thaliana*. The Journal of Biological Chemistry. 2001;**276**(28):26405-26410

[53] Inoue T, Higuchi M, Hashimoto Y, Seki M, Kobayashi M, Kato T, et al. Identification of CRE1 as a cytokinin receptor from *Arabidopsis*. Nature. 2001;**409**(6823):1060

[54] Hwang I, Sheen J. Two-component circuitry in *Arabidopsis* cytokinin signal transduction. Nature. 2001;**413**(6854):383

[55] To JP, Kieber JJ. Cytokinin signaling: Two-components and more. Trends in Plant Science. 2008;**13**(2):85-92

[56] Suzuki T, Imamura A, Ueguchi C, Mizuno T. Histidine-containing phosphotransfer (HPt) signal transducers implicated in His-to-Asp phosphorelay in *Arabidopsis*. Plant & Cell Physiology. 1998;**39**(12):1258-1268

[57] Tanaka Y, Suzuki T, Yamashino T, MIZUNO T. Comparative studies of the AHP histidine-containing phosphotransmitters implicated in His-to-Asp phosphorelay in *Arabidopsis thaliana*. Bioscience, Biotechnology, and Biochemistry. 2004;**68**(2):462-465

[58] Hutchison CE, Li J, Argueso C, Gonzalez M, Lee E, Lewis MW, et al. The *Arabidopsis* histidine phosphotransfer proteins are redundant positive regulators of cytokinin signaling. The Plant Cell. 2006;**18**(11):3073-3087

[59] To JP, Haberer G, Ferreira FJ, Deruere J, Mason MG, Schaller GE, et al. Type-A *Arabidopsis* response regulators are partially redundant negative regulators of cytokinin signaling. The Plant Cell. 2004;**16**(3):658-671

[60] Mason MG, Mathews DE, Argyros DA, Maxwell BB, Kieber JJ, Alonso JM, et al. Multiple type-B response regulators mediate cytokinin signal transduction in *Arabidopsis*. The Plant Cell. 2005;**17**(11):3007-3018

[61] Sakai H, Honma T, Aoyama T, Sato S, Kato T, Tabata S, et al. ARR1, a transcription factor for genes immediately responsive to cytokinins. Science. 2001;**294**(5546):1519-1521

[62] Müller B, Sheen J. Cytokinin and auxin interaction in root stem-cell specification during early embryogenesis. Nature. 2008;**453**(7198):1094

[63] Faiss M, Zalubìlová J, Strnad M, Schmülling T. Conditional transgenic expression of the ipt gene indicates a function for cytokinins in paracrine signaling in whole tobacco plants. The Plant Journal. 1997;**12**(2):401-415

[64] Ioio RD, Linhares FS, Scacchi E, Casamitjana-Martinez E, Heidstra R, Costantino P, et al. Cytokinins determine *Arabidopsis* root-meristem size by controlling cell differentiation. Current Biology. 2007;**17**(8):678-682

[65] Matsumoto-Kitano M, Kusumoto T, Tarkowski P, Kinoshita-Tsujimura K,

Václavíková K, Miyawaki K, et al. Cytokinins are central regulators of cambial activity. Proceedings of the National Academy of Sciences. 2008;**105**(50):20027-20031

[66] Bürkle L, Cedzich A, Döpke C, Stransky H, Okumoto S, Gillissen B, et al. Transport of cytokinins mediated by purine transporters of the PUP family expressed in phloem, hydathodes, and pollen of *Arabidopsis*. The Plant Journal. 2003;**34**(1):13-26

[67] Sun J, Hirose N, Wang X, Wen P, Xue L, Sakakibara H, et al. *Arabidopsis* SOI33/AtENT8 gene encodes a putative equilibrative nucleoside transporter that is involved in cytokinin transport in planta. Journal of Integrative Plant Biology. 2005;**47**(5):588-603

[68] Hirose N, Takei K, Kuroha T, Kamada-Nobusada T, Hayashi H, Sakakibara H. Regulation of cytokinin biosynthesis, compartmentalization and translocation. Journal of Experimental Botany. 2007;**59**(1):75-83

[69] Ko D, Kang J, Kiba T, Park J, Kojima M, Do J, et al. *Arabidopsis* ABCG14 is essential for the root-to-shoot translocation of cytokinin. Proceedings of the National Academy of Sciences. 2014;**111**(19):7150-7155. DOI: 10.1073/pnas.1321519111

[70] Zhang K, Novak O, Wei Z, Gou M, Zhang X, Yu Y, et al. *Arabidopsis* ABCG14 protein controls the acropetal translocation of root-synthesized cytokinins. Nature Communications. 2014;**5**:3274

[71] Ioio RD, Nakamura K, Moubayidin L, Perilli S, Taniguchi M, Morita MT, et al. A genetic framework for the control of cell division and differentiation in the root meristem. Science. 2008;**322**(5906):1380-1384

[72] Beemster GT, Baskin TI. STUNTED PLANT 1 mediates effects of cytokinin, but not of auxin, on cell division and expansion in the root of *Arabidopsis*. Plant Physiology. 2000;**124**(4):1718-1727

[73] Blilou I, Xu J, Wildwater M, Willemsen V, Paponov I, Friml J, et al. The PIN auxin efflux facilitator network controls growth and patterning in *Arabidopsis* roots. Nature. 2005;**433**(7021):39

[74] Moubayidin L, Di Mambro R, Sozzani R, Pacifici E, Salvi E, Terpstra I, et al. Spatial coordination between stem cell activity and cell differentiation in the root meristem. Developmental Cell. 2013;**26**(4):405-415

[75] Růžička K, Šimášková M, Duclercq J, Petrášek J, Zažímalová E, Simon S, et al. Cytokinin regulates root meristem activity via modulation of the polar auxin transport. Proceedings of the National Academy of Sciences. 2009;**106**(11):4284-4289

[76] Pernisová M, Klíma P, Horák J, Válková M, Malbeck J, Souček P, et al. Cytokinins modulate auxin-induced organogenesis in plants via regulation of the auxin efflux. Proceedings of the National Academy of Sciences. 2009;**106**(9):3609-3614

[77] Laplaze L, Benkova E, Casimiro I, Maes L, Vanneste S, Swarup R, et al. Cytokinins act directly on lateral root founder cells to inhibit root initiation. The Plant Cell. 2007;**19**(12):3889-3900

[78] Riefler M, Novak O, Strnad M, Schmülling T. *Arabidopsis* cytokinin receptor mutants reveal functions in shoot growth, leaf senescence, seed size, germination, root development, and cytokinin metabolism. The Plant Cell. 2006;**18**(1):40-54

[79] Kuderová A, Urbánková I, Válková M, Malbeck J, Brzobohatý B, Nemethova D, et al. Effects of conditional IPT-dependent cytokinin overproduction on root architecture

of *Arabidopsis* seedlings. Plant & Cell Physiology. 2008;**49**(4):570-582

[80] Li X, Mo X, Shou H, Wu P. Cytokinin-mediated cell cycling arrest of pericycle founder cells in lateral root initiation of *Arabidopsis*. Plant & Cell Physiology. 2006;**47**(8):1112-1123

[81] Chang L, Ramireddy E, Schmülling T. Lateral root formation and growth of *Arabidopsis* is redundantly regulated by cytokinin metabolism and signalling genes. Journal of Experimental Botany. 2013;**64**(16):5021-5032

[82] Zhao Y, Cheng S, Song Y, Huang Y, Zhou S, Liu X, et al. The interaction between rice ERF3 and WOX11 promotes crown root development by regulating gene expression involved in cytokinin signaling. The Plant Cell. 2015;**27**(9):2469-2483. DOI: 10.1105/tpc.15.00227

[83] Jeon J, Cho C, Lee MR, Van Binh N, Kim J. CYTOKININ RESPONSE FACTOR2 (CRF2) and CRF3 regulate lateral root development and response to cold stress in *Arabidopsis*. The Plant Cell. 2016;**28**(8):1828-1843. DOI: 10.1105/tpc.15.00909

[84] Signora L, De Smet I, Foyer CH, Zhang H. ABA plays a central role in mediating the regulatory effects of nitrate on root branching in *Arabidopsis*. The Plant Journal. 2001;**28**(6):655-662

[85] Brady SM, Sarkar SF, Bonetta D, McCourt P. The ABSCISIC ACID INSENSITIVE 3 (ABI3) gene is modulated by farnesylation and is involved in auxin signaling and lateral root development in *Arabidopsis*. The Plant Journal. 2003;**34**(1):67-75

[86] De Smet I, Signora L, Beeckman T, Inzé D, Foyer CH, Zhang H. An abscisic acid-sensitive checkpoint in lateral root development of *Arabidopsis*. The Plant Journal. 2003;**33**(3):543-555

[87] Shkolnik-Inbar D, Bar-Zvi D. ABI4 mediates abscisic acid and cytokinin inhibition of lateral root formation by reducing polar auxin transport in *Arabidopsis*. The Plant Cell. 2010;**22**(11):3560-3573. DOI: 10.1105/tpc.110.074641

[88] Ding ZJ, Yan JY, Li CX, Li GX, Wu YR, Zheng SJ. Transcription factor WRKY 46 modulates the development of *Arabidopsis* lateral roots in osmotic/salt stress conditions via regulation of ABA signaling and auxin homeostasis. The Plant Journal. 2015;**84**(1):56-69

[89] Hansen J. Light dependent promotion and inhibition of adventitious root formation by gibberellic acid. Planta. 1975;**123**(2):203-205

[90] Yaxley JR, Ross JJ, Sherriff LJ, Reid JB. Gibberellin biosynthesis mutations and root development in pea. Plant Physiology. 2001;**125**(2):627-633

[91] Fu X, Harberd NP. Auxin promotes *Arabidopsis* root growth by modulating gibberellin response. Nature. 2003;**421**(6924):740

[92] Steffens B, Wang J, Sauter M. Interactions between ethylene, gibberellin and abscisic acid regulate emergence and growth rate of adventitious roots in deepwater rice. Planta. 2006;**223**(3):604-612

[93] Müssig C, Shin G-H, Altmann T. Brassinosteroids promote root growth in *Arabidopsis*. Plant Physiology. 2003;**133**(3):1261-1271

[94] Bao F, Shen J, Brady SR, Muday GK, Asami T, Yang Z. Brassinosteroids interact with auxin to promote lateral root development in *Arabidopsis*. Plant Physiology. 2004;**134**(4):1624-1631

[95] Raya-González J, Pelagio-Flores R, López-Bucio J. The jasmonate receptor COI1 plays a role in jasmonate-induced

lateral root formation and lateral root positioning in *Arabidopsis thaliana*. Journal of Plant Physiology. 2012;**169**(14):1348-1358

[96] Hentrich M, Böttcher C, Düchting P, Cheng Y, Zhao Y, Berkowitz O, et al. The jasmonic acid signaling pathway is linked to auxin homeostasis through the modulation of YUCCA 8 and YUCCA 9 gene expression. The Plant Journal. 2013;**74**(4):626-637

[97] Cai X-T, Xu P, Zhao P-X, Liu R, Yu L-H, Xiang C-B. *Arabidopsis* ERF109 mediates cross-talk between jasmonic acid and auxin biosynthesis during lateral root formation. Nature Communications. 2014;**5**:5833

[98] Gutierrez L, Mongelard G, Floková K, Păcurar DI, Novák O, Staswick P, et al. Auxin controls *Arabidopsis* adventitious root initiation by regulating jasmonic acid homeostasis. The Plant Cell. 2012;**24**(6):2515-2527. DOI: 10.1105/tpc.112.099119

[99] Strader LC, Chen GL, Bartel B. Ethylene directs auxin to control root cell expansion. The Plant Journal. 2010;**64**(5):874-884

[100] Ivanchenko MG, Muday GK, Dubrovsky JG. Ethylene–auxin interactions regulate lateral root initiation and emergence in *Arabidopsis thaliana*. The Plant Journal. 2008;**55**(2):335-347

[101] Lewis DR, Negi S, Sukumar P, Muday GK. Ethylene inhibits lateral root development, increases IAA transport and expression of PIN3 and PIN7 auxin efflux carriers. Development. 2011;**138**(16):3485-3495. DOI: 10.1242/dev.065102

[102] Jiang L, Matthys C, Marquez-Garcia B, De Cuyper C, Smet L, De Keyser A, et al. Strigolactones spatially influence lateral root development through the cytokinin signaling network. Journal of Experimental Botany. 2015;**67**(1):379-389

Chapter 3

Cross Talk among Phytohormone Signal and Carbohydrate Metabolism Involving Regenerable Calli Induction under Osmotic Treatment

Hsiang-Ting Lee and Wen-Lii Huang

Abstract

Nonregenerable calli (NRC) derived from immature seeds of japonica rice were inoculated on MS medium containing 10 μM 2,4-D (MSD10). They turned to highly regenerable calli (HRC) when sorbitol was supplemented into the medium. Meanwhile, high levels of endogenous IAA and ABA were accumulated in HRC. Exogenous IAA precursor and ABA in MSD10 have the same effect to enhance regeneration ability. However, there are only partial effects if IAA precursor or ABA was supplemented, respectively. The regeneration ability is prominently decreased from 75% to 25% while an auxin transport inhibitor, 2,3,5-triiodobenzoic acid, was included in the medium. It suggested that endogenous auxin signal and ABA may involve in the induction of HRC. Furthermore, it showed higher contents of glucose, sucrose, and starch and higher expression levels of wall-bound invertase 1, sucrose transporter 1 (OsSUT1), and OsSUT2 genes in HRC than in NRC. The expression levels of PIN-formed 1 and LEA1 were also consistent with the trend of carbohydrate metabolisms. We thus concluded a flowchart for HRC induction by osmotic stress. According to the hypothesis, osmotic stress may regulate endogenous levels of auxin interacting with ABA, then affect carbohydrate metabolism to trigger callus initiation and further shoot regeneration in rice.

Keywords: phytohormone, osmotic stress, carbohydrate metabolism, plant regeneration, rice

1. Introduction

Totipotency in plant cells allow them to differentiate and regenerate from one single cell into whole plants under conditioned culture [1]. Many factors, i.e., cultivars, carbon sources, phytohormones, and osmotic stress, affect the cell totipotency (**Figure 1**). So far, the regeneration cultures have been successfully established in many plant species including rice, but molecular mechanisms behind this scene are still lots of mist [2–4]. In rice, pluripotent calli were usually induced from seeds or immature embryos cultured on MS medium containing

2,4-D. These calli derived from various cultivars can be classified into two different cell types, nonregenerable callus (NRC) and highly regenerable callus (HRC). After transferred to regeneration medium, HRC can rapidly process shoot organogenesis, but NRC will still retain callus amplification or adventitious root formation [5]. There have been many protocols developed to optimize the induction frequency of NRC, but the shoot regeneration ability still varies among cultivars [6–8].

Previous studies had identified that added osmotic agents like sorbitol or mannitol in callus induction medium can stimulate HRC formation instead of NRC, thus promoting the shoot formation frequency [5, 7, 9–11]. It is still unclear why appropriate osmotic stress during callus induction can promote shoot organogenesis frequency, but there were lots of studies indicating that osmotic stress can stimulate endogenous phytohormone abscisic acid (ABA) accumulation which is also proven to have the function of promoting somatic embryogenesis and shoot organogenesis when it used as an exogenous plant growth regulator in callus culture [11–15]. ABA is widely recognized as a negative plant hormone which mainly functions on stress responses and seed dormancy, but when it is treated in low concentration, ABA could become a positive regulator on root elongation [16]. Although the molecular mechanisms behind this phenomenon are not yet clarified, apparently, ABA has its role in the developmental process.

Other phytohormones, auxin and cytokinin, are also reported to play critical roles in embryogenic callus induction and shoot development [8, 17, 18]. In plants, auxin is considered as a positive growth regulator, which contributes to cell differentiation [19]. High levels of auxin are usually found in shoot and root apical meristems where plants organize their developmental patterns, and so does in HRC [20, 21]. Many studies mentioned that adding different levels of exogenous auxin during callus culture may cause different organogenesis, like low levels of auxin may induce root organ formation while high levels of 2,4-D sustain callus induction [22–24]. Cross talk among auxin and ABA mostly focused on root development or abiotic stress responses [25–30]. Our previous studies had identified that both endogenous auxin and ABA levels were higher in HRC than in NRC and then quickly decreased after transferred to regeneration medium [5, 8]. The interaction and signaling pathway between these two phytohormones to shoot organogenesis are further discussed.

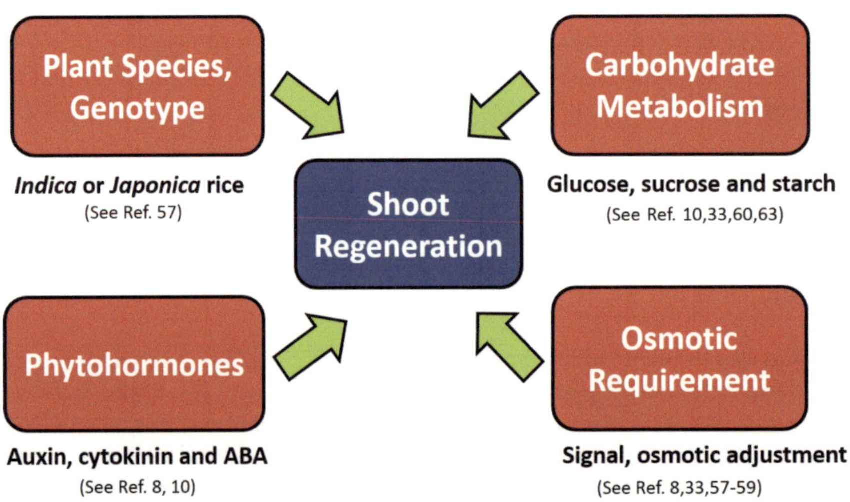

Figure 1.
Factors affecting shoot regeneration in rice.

Undoubtedly, shoot regeneration is a very complicated process, which is modulated by networks among phytohormones and their downstream metabolic changes in plant cells [31, 32]. In our previous studies, we reported that HRC contained higher levels of glucose, sucrose, and starch, accomplished with higher mRNA transcription levels of cell-wall bound-invertase 1 (CIN1) and SUTs genes [8, 33]. Also, we found that these changes on carbohydrate metabolisms can also be found in osmotic and ABA treated calli, suggesting the possibility of regulation mechanisms on HRC formation [5, 11].

Combined with our findings and previous studies, we proposed a hypothesis for HRC induction in rice callus culture. Osmotic stress may initiate a signal to regulate endogenous auxin to interact with ABA and then stimulate soluble sugars accumulate in callus to trigger embryonic callus formation for shoot regeneration [11].

2. Effect of phytohormones on shoot regeneration in rice callus culture

Phytohormones are considered one of the major important factors that control cell fate in callus cultures. Different types and concentrations of auxin treatment may induce different tissue differentiations. Besides, ABA is also considered to play roles in somatic embryogenesis and shoot organogenesis. In this section, we will discuss the functions of auxin and ABA in regenerable callus induction.

2.1 Roles of auxin in regenerable callus induction

Auxin is known to play a major role in cell elongation, growth tropisms, and cell fate determination [34]. The main endogenous auxin compound is indole-3-acetic acid (IAA), which could be synthesized via a tryptophan-dependent or -independent pathway [35]. After being synthesized in apical meristems, auxin will be transported to the other tissues through its transporter, PIN-formed proteins (PIN), and AUXIN1/LIKE-AUX1 (AUX/LAX) proteins to deliver the hormonal signal for downstream auxin responses [17, 36]. The distribution of auxin in plants is polarity which means that it is not equally expressed in the whole plant but specifically concentrated in some tissues or limited to cells [36]. Due to this property, scientists are able to monitor the patterns of auxin gradient during plant development [37]. Form the literature, the positions of maximum auxin accumulation were the place processing organ initiation. Therefore, auxin is usually found to have maximum levels in apical meristems or in the developing tissues [38]. During embryogenesis, the expression pattern of PINs dynamically changes within the developmental stages. For example, PIN1 is expressed without polarity until 16-cell stage, but later in 32-cell stage, it will express specifically in the basal part of provascular cells to direct auxin flow to hypophysis. After dividing into transition and early heart embryo stages, the expression of PIN1 will shift to the flank of the apical part, thus accumulating auxin in the edge of apical domain to shape the embryo [39, 40]. Similar developmental patterns can be found during shoot apical meristem (SAM) establishment, where auxin is highly concentrated in leaf primordia and the central region of SAM [18, 41].

In callus, the signal of auxin is mainly located in the central region near callus induction medium. Later, the signal will shift to the surface layer and start SAM establishment for shoot differentiation after transfer into regeneration medium [42]. However, this organogenesis process is blocked when the activity of auxin transporters is inhibited, and so does in auxin sensor *shoot meristemless* (*stm*) mutants [43]. In our previous studies, we compared endogenous IAA levels between NRC and HRC in rice callus and found that HRC has higher IAA content than NRC, and so does the mRNA expression levels of PIN1, suggesting that HRC has higher auxin sensitivity, which results in higher regeneration ability [5, 8, 11]. Also, there are researches mentioning that overexpression of STM can maintain the somatic embryogenesis frequency even under low concentrations of 2,4-D [44]. Besides, increasing the expression levels of auxin biosynthesis regulator *YUCCA* (*YUC*) genes can also promote shoot organogenesis ability in callus [45]. Similar results can be found in supplement of IAA precursor anthranilic acid (An) in rice callus culture; the shoot regeneration frequency increased by 35% under An treatment, suggesting that a high endogenous auxin level is required for HRC formation [5].

2.2 Roles of ABA on regenerable callus induction

ABA is usually considered to play negative roles in plant growth [26, 46]. Except the function of seed dormancy regulation and stress responses, ABA is also reported to have function on root and shoot development [46, 47]. Although there were some genes participating in ABA signaling, which were also reported to involve in the shoot regeneration process, there is still no clarity about the molecular function of endogenous ABA on the shoot organogenesis process [14, 15]. However, some studies, including our previous works, found that ABA was highly accumulated in HRC, but not in NRC [5]. Furthermore, the expression profiles of ABA biosynthesis genes were also upregulated, which match with our results [48].

On the other hand, ABA is reported to stimulate dehydration responses during embryo maturation stages [49]. From the publications, we observed that HRC has less water content and smaller callus size, which is similar to the dehydration phenotype. Thus, it is possible that ABA shared similar regulation mechanisms with embryo maturation during shoot regeneration. Despite the loss of water in HRC, it also showed higher content of soluble sugars [8, 11]. It is already known that high content of sugar may enhance osmotic stress and then stimulate endogenous ABA biosynthesis to ABA responses [33], but the underlying mechanisms to shoot differentiation is still unclarified.

3. Cross talk among osmotic stress and phytohormones in callus culture

In cells, osmotic stress could originate from water-deficiency or high salt which caused an imbalance between plastids and apoplast [50]. To achieve tolerant to osmotic stress condition, cells will modulate the content's osmotic adjustments like sugars, potassium ions, or proline to balance its osmotic pressure to environments to avoid collapse. Phytohormone ABA is reported to accumulate under osmotic stress to modulate anion channel *SLOW ANION CHANNEL-ASSOCIATED1* (*SLAC1*) to stomata closure and the potassium transporter KUP to potassium homeostasis [9], while in plant tissue culture, appropriate osmotic stress can help embryonic callus formation [7]. In our case, we found that HRC showed dosage responses to osmotic treatments and has the highest induction frequency under 0.6 M sorbitol treatment. We also noticed that osmotic requirement is various among rice species, some cultivars require higher osmotic stress to induce HRC and some require lower, and even one cultivar can form HRC rapidly without osmotic treatments. We then analyzed the ABA response in those calli and found that HRC does have higher *LATE EMBRYOGENESIS ABOUNDENCE 1* (*LEA1*) gene expressions, which is commonly used as ABA signaling marker. Interestingly, *LEA1* is also upregulated in the rice cultivar without osmotic stress treatment, indicated the effect of osmotic stress could be on stimulate ABA biosynthesis and its downstream responses during embryonic callus induction [5, 11].

Not only ABA, auxin is also regulated by osmotic stress. In rice, the endogenous levels of auxin are reported to be suppressed under osmotic stress [50, 51]. However, a closer look at the expression levels of different auxin biosynthesis-related genes and the distributions of auxin showed various patterns in the whole plants in different stages, some of them even inconsistent with the overall patterns, suggesting that auxin may function differently among tissues under osmotic stress. As for the patterns of *PINs* under osmotic stress, one of the researches reported that osmotic stress may inhibit the expression of *PIN1* in leaf primordia, thus suppressing shoot development [52]. However, our previous works indicated that *OsPIN1* could be upregulated by 0.6 M sorbitol treatment in HRC, and also in the nonosmotic requirement cultivars [11]. Although there have been many studies performing transcriptomic or proteomic studies of auxin responses under osmotic stress [53, 54], how osmotic stress directs with auxin responses to HRC formation is still less known.

4. Roles of carbohydrate metabolisms during HRC induction under osmotic stress treatment

Exogenous carbohydrates are used either as energy sources or osmotic agents in tissue culture. So far, many articles have revealed that different carbon sources may lead to different callus induction abilities [7, 55, 56], but rarely discussed about the carbohydrate metabolisms and signaling pathways in callus cultures. Sucrose is widely used as a main carbon source as well as osmotic agent in rice tissue culture [57, 58]. During the tissue culture, sucrose will be taken up and hydrolyzed into glucose and fructose by CINs, or be transported into cells by SUTs for further application [59]. CINs were already reported to involve in early seeding establishments and grain fillings, and so do the SUTs [60, 61].

According to our studies, HRC induced by osmotic stress seemed to obtain higher contents of glucose, sucrose, and starch than NRC, which also reflected in the higher dry weights [33, 56, 62]. We analyzed the expression patterns of *CIN1* and *SUTs* during rice callus culture. The expression of *CIN1*, *SUT1*, and *SUT2* in HRC was upregulated by osmotic stress, but not in NRC, while in the nonosmotic requirement cultivars, there of these sucrose-uptake genes were expressed earlier than in the cultivar of low-regenerable ability [8, 11]. It is suggested that higher soluble sugars in HRC may be caused from higher sucrose uptake and translocation under stress treatment. Besides, we also observed that osmotic stress induced HRC has lower α-amylase activities and thus increases the content of starches [33], while in nonosmotic requirement cultivar, the callus rather induced ADPG pyrophosphorylase (AGPase) activity to accumulate starches [62]. The results suggested two different regulations on carbohydrate metabolisms to HRC induction. Although there are different ways of starch accumulation in HRC, the accumulated carbohydrates will soon degrade after transplanting the callus onto regeneration medium in 3 days, suggesting that these carbohydrates could be stored as a carbon source in HRC and then used for the developmental process during the regeneration stages [8, 11, 33].

Plants are known to accumulate starch granules specifically in the columella cells [63, 64]. Although the function of these starch granules is mostly reported in root gravity, they could also be markers to point out the stem cell niche, since these starch granules may disappear in the plants with stem cell defect [64]. Our studies also found the accumulation of starch granules in peripheral regions in HRC [62].

High concentration of sorbitol or mannitol will enlarge the distribution of starch granules. It may link to the increase of the shoot organogenesis area [5]. However, the physiological functions and mechanisms of starch accumulation still remained, requiring further studies.

Moreover, the accumulation and metabolism of soluble sugars and starch can be induced by AnA treatment (anthranilic acid and ABA supplemented into the medium together) to replace osmotic stress treatment. High levels of endogenous IAA and ABA at the same time are necessary during HRC induction. Both of them need to decreased suddenly in few days are also an important criteria for further shoot regeneration [5]. To link these metabolic changes with phytohormone regulations, we also introduced auxin transport inhibitor TIBA during callus induction and found that it will inhibit carbohydrate accumulation and result in low shoot regeneration frequency. However, ABA signals seemed to be promoted under TIBA treatment [8, 11]. It is still not clear whether exceeding ABA signals will turn into negative regulator on HRC induction, but these results still indicate that there must be interactions among auxin, ABA, and carbohydrate metabolisms on HRC formation.

5. Conclusions

Inducing regenerate tissues from pluripotent cells is a fascinating event. So far, botanists have already shown that they were able to get regeneration plants from callus in many plant species [49, 65–67]. However, why and how plants achieve this process is still unknown, especially in molecular levels. Here, we propose a hypothesis among phytohormone, osmotic stress, and carbohydrate metabolisms on HRC induction based on current knowledge and our findings (**Figure 2**). According to our model, levels of endogenous IAA upregulated by osmotic stress treatment can promote sugar uptake via CIN and SUT, which result in carbohydrate accumulation during callus induction stages. Similar to auxin, endogenous ABA level is also enhanced under osmotic stress, thus modulating starch accumulation during formation of HRC by downregulating α-amylase activity. Our studies indicated that exogenous auxin or ABA treatment alone is not sufficient for embryonic or organogenic callus formation, which only increased the plant regeneration rate for 35 and 5%, separately [5]. However, when we combine both ABA and anthranilic acid treatment together, the regeneration frequency can be promoted to 80% similar to osmotic stress treatment, suggesting that there must be some interaction between these two phytohormones. The roles of accumulated carbohydrates in HRC could be used as osmotic agents for further metabolism changes or be consumed as an energy

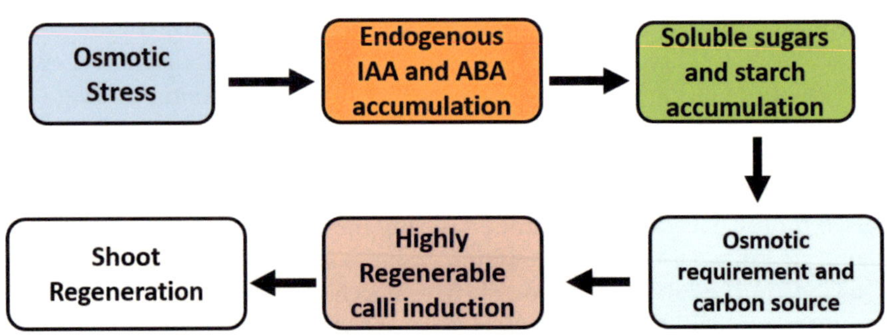

Figure 2.
Working hypothesis of highly regenerable callus induction under osmotic stress treatment in rice.

source in later regeneration stages. In conclusion, the culture system of shoot regeneration in rice callus is a two-step process. Our studies suggested that induction of highly regenerable callus is more important than different kinds of treatment during the shoot regeneration stage. Besides, osmotic stress triggers a serial of change of endogenous hormone metabolism, sensing, and signal transduction, which leads to increase of sucrose uptake and starch accumulation and provides sufficient carbon source for further shoot regeneration.

Acknowledgements

The authors would like to thank Professor Li-Fei Liu, Department of Agronomy, National Taiwan University, to establish the high efficiency of the shoot regeneration system by osmotic stress treatment in rice callus culture. We would like to thank the Council of Agriculture, Taiwan, ROC, for providing the funding to finish this study.

Conflict of interest

The authors have no conflict of interest.

Author details

Hsiang-Ting Lee[1] and Wen-Lii Huang[2]*

1 Institute of Molecular Biology, Academia Sinica, Taipei, Taiwan

2 Department of Agronomy, National Chiayi University, Chiayi, Taiwan

*Address all correspondence to: wlhuang@mail.ncyu.edu.tw

References

[1] Reinert J. Über die Kontrolle der Morphogenese und die Induktion von Adventivembryonen an Gewebekulturen aus Karotten. Planta. 1959;**53**:318-333

[2] Indoliya Y, Tiwari P, Chauhan AS, Goel R, Shri M, Bag SK, et al. Decoding regulatory landscape of somatic embryogenesis reveals differential regulatory networks between japonica and indica rice subspecies. Scientific Reports. 2016;**6**:23050. DOI: 10.1038/srep23050

[3] Hoque M, Mansfield JW. Effect of genotype and explant age on callus induction and subsequent plant regeneration from root-derived callus of indica rice genotypes. Plant Cell, Tissue and Organ Culture. 2004;**78**:217223. DOI: 10.1023/b:ticu.0000025640.75168.2d

[4] Liu C, Xia X, Yin W, Huang L, Zhou J. Shoot regeneration and somatic embryogenesis from needles of redwood (*Sequoia sempervirens* (D. Don.) Endl.). Plant Cell Reports. 2006;**25**:621628. DOI: 10.1007/s00299-006-0120-y

[5] Huang W-L, Lee C-H, Chen Y-R. Levels of endogenous abscisic acid and indole-3-acetic acid influence shoot organogenesis in callus cultures of rice subjected to osmotic stress. Plant Cell, Tissue and Organ Culture (PCTOC). 2011;**108**:257263. DOI: 10.1007/s11240-011-0038-0

[6] Al-Khayri JM, Shamblin CE, Anderson EJ. Callus induction and plant regeneration of U.S. rice genotypes as affected by medium constituents. In Vitro Cellular & Developmental Biology—Plant. 1996;**32**:227232. DOI: 10.1007/bf02822692

[7] Jain RK, Davey MR, Cocking EC, Wu R. Carbohydrate and osmotic requirements for high-frequency plant regeneration from protoplast-derived colonies of indica and japonica rice varieties. Journal of Experimental Botany. 1997;**48**:751-758. DOI: 10.1093/jxb/48.3.751

[8] Lee S-T, Huang W-L. Cytokinin, auxin, and abscisic acid affects sucrose metabolism conduce to de novo shoot organogenesis in rice (*Oryza sativa* L.) callus. Botanical Studies. 2013;**54**:5. DOI: 10.1186/1999-3110-54-5

[9] Wilson ME, Mixdorf M, Berg R, Haswell ES. Plastid osmotic stress influences cell differentiation at the plant shoot apex. Development. 2016;**143**:3382-3393. DOI: 10.1242/dev.136234

[10] Feng X, Zhao P, Hao J, Hu J, Kang D, Wang H. Effects of sorbitol on expression of genes involved in regeneration of upland rice (*Oryza sativa* L.). Plant Cell, Tissue and Organ Culture (PCTOC). 2011;**106**:455463. DOI: 10.1007/s11240-011-9943-5

[11] Lee S-T, Huang W-L. Osmotic stress stimulates shoot organogenesis in callus of rice (*Oryza sativa* L.) via auxin signaling and carbohydrate metabolism regulation. Plant Growth Regulation. 2013;**73**:193204. DOI: 10.1007/s10725-013-9880-x

[12] Elmaghrabi AM, Rogers HJ, Francis D, Ochatt SJ. PEG Induces high expression of the cell cycle checkpoint gene WEE1 in embryogenic callus of Medicago truncatula: Potential link between cell cycle checkpoint regulation and osmotic stress. Frontiers in Plant Science. 2017;**8**:1479. DOI: 10.3389/fpls.2017.01479

[13] Sholi NJ, Chaurasia A, Agrawal A, Sarin N. ABA enhances plant regeneration of somatic embryos

derived from cell suspension cultures of plantain cv. Spambia (Musa sp.). Plant Cell, Tissue and Organ Culture (PCTOC). 2009;**99**:133140. DOI: 10.1007/s11240-009-9585-z

[14] Guillou C, Fillodeau A, Brulard E, Breton D, Maraschin S, Verdier D, et al. Indirect somatic embryogenesis of *Theobroma cacao* L. In liquid medium and improvement of embryo-to-plantlet conversion rate. In Vitro Cellular & Developmental Biology—Plant. 2018;**54**:377-391. DOI: 10.1007/s11627-018-9909-y

[15] Rai MK, Shekhawat N, Harish GAK, Phulwaria M, Ram K, et al. The role of abscisic acid in plant tissue culture: A review of recent progress. Plant Cell, Tissue and Organ Culture (PCTOC). 2011;**106**:179190. DOI: 10.1007/s11240-011-9923-9

[16] Chen Y-S, Lo S-F, Sun P-K, Lu C-A, Ho T-HD, Yu S-M. A late embryogenesis abundant protein HVA1 regulated by an inducible promoter enhances root growth and abiotic stress tolerance in rice without yield penalty. Plant Biotechnology Journal. 2015;**13**:105-116. DOI: 10.1111/pbi.12241

[17] Yamaguchi N, Wu M-F, Winter CM, Berns MC, Nole-Wilson S, Yamaguchi A, et al. A Molecular framework for auxin-mediated initiation of flower primordia. Developmental Cell. 2013;**24**:271-282. DOI: 10.1016/j.devcel.2012.12.017

[18] Sassi M, Vernoux T. Auxin and self-organization at the shoot apical meristem. Journal of Experimental Botany. 2013;**64**:2579-2592. DOI: 10.1093/jxb/ert101

[19] Xu C, Cao H, Zhang Q, Wang H, Xin W, Xu E, et al. Control of auxin-induced callus formation by bZIP59-LBD complex in Arabidopsis regeneration. Nature Plants. 2018. DOI: 10.1038/s41477-017-0095-4

[20] Xu J, Hofhuis H, Heidstra R, Sauer M, Friml J, Scheres B. A molecular framework for plant regeneration. Science. 2006;**311**:385-388. DOI: 10.1126/science.1121790

[21] Shin J, Seo PJ. Varying auxin levels induce distinct pluripotent states in callus cells. Frontiers in Plant Science. 2018;**9**:1653. DOI: 10.3389/fpls.2018.01653

[22] Pernisová M, Klíma P, Horák J, Válková M, Malbeck J, Souček P, et al. Cytokinins modulate auxin-induced organogenesis in plants via regulation of the auxin efflux. Proceedings of the National Academy of Sciences. 2009;**106**:3609-3614. DOI: 10.1073/pnas.0811539106

[23] Vanneste S, Friml J. Auxin: A trigger for change in plant development. Cell. 2009;**136**:1005-1016. DOI: 10.1016/j.cell.2009.03.001

[24] Coudert Y, Périn C, Courtois B, Khong N, Gantet P. Genetic control of root development in rice, the model cereal. Trends in Plant Science. 2010;**15**:219-226. DOI: 10.1016/j.tplants.2010.01.008

[25] Cline MG, Choonseok O. A reappraisal of the role of abscisic acid and its interaction with auxin in apical dominance. Annals of Botany. 2006;**98**:891-897. DOI: 10.1093/aob/mcl173

[26] Belin C, Megies C, Hauserová E, Lopez-Molina L. Abscisic acid represses growth of the Arabidopsis embryonic axis after germination by enhancing auxin signaling. The Plant Cell. 2009;**21**:2253-2268. DOI: 10.1105/tpc.109.067702

[27] Popko J, Hänsch R, Mendel R-R, Polle A, Teichmann T. The role of abscisic acid and auxin in the response of poplar to abiotic stress.

Plant Biology. 2010;**12**:242-258. DOI: 10.1111/j.1438-8677.2009.00305.x

[28] Anderson JV, Doğramacı M, Horvath DP, Foley ME, Chao WS, Suttle JC, et al. Auxin and ABA act as central regulators of developmental networks associated with paradormancy in Canada thistle (*Cirsium arvense*). Functional & Integrative Genomics. 2012;**12**:515-531. DOI: 10.1007/s10142-012-0280-5

[29] Zörb C, Geilfus C-M, Mühling KH, Ludwig-Müller J. The influence of salt stress on ABA and auxin concentrations in two maize cultivars differing in salt resistance. Journal of Plant Physiology. 2013;**170**:220-224. DOI: 10.1016/j.jplph.2012.09.012

[30] Xu W, Jia L, Shi W, Liang J, Zhou F, Li Q , et al. Abscisic acid accumulation modulates auxin transport in the root tip to enhance proton secretion for maintaining root growth under moderate water stress. The New Phytologist. 2013;**197**:139-150. DOI: 10.1111/nph.12004

[31] Perez-Garcia P, Moreno-Risueno MA. Stem cells and plant regeneration. Developmental Biology. 2018;**442**:3-12. DOI: 10.1016/j.ydbio.2018.06.021

[32] Kareem A, Durgaprasad K, Sugimoto K, Du Y, Pulianmackal AJ, Trivedi ZB, et al. PLETHORA genes control regeneration by a two-step mechanism. Current Biology: CB. 2015;**25**:1017-1030. DOI: 10.1016/j.cub.2015.02.022

[33] Huang W-L, Liu L-F. Carbohydrate metabolism in rice during callus induction and shoot regeneration induced by osmotic stress. Botanical Bulletin of Academia Sinica. 2002;**43**

[34] Saini S, Sharma I, Kaur N, Pati P. Auxin: A master regulator in plant root development. Plant Cell Reports. 2013;**32**:741-757. DOI: 10.1007/s00299-013-1430-5

[35] Zhang Z, Tucker E, Hermann M, Laux T. A Molecular framework for the embryonic initiation of shoot meristem stem cells. Developmental Cell. 2017;**40**:264-277.e4. DOI: 10.1016/j.devcel.2017.01.002

[36] Brumos J, Robles LM, Yun J, Vu TC, Jackson S, Alonso JM, et al. Local auxin biosynthesis is a key regulator of plant development. Developmental Cell. 2018;**47**:306-318.e5. DOI: 10.1016/j.devcel.2018.09.022

[37] Liao C-YY, Smet W, Brunoud G, Yoshida S, Vernoux T, Weijers D. Reporters for sensitive and quantitative measurement of auxin response. Nature Methods. 2015;**12**: 207-210. DOI: 10.1038/nmeth.3279

[38] Marhava P, Bassukas A, Zourelidou M, Kolb M, Moret B, Fastner A, et al. A molecular rheostat adjusts auxin flux to promote root protophloem differentiation. Nature. 2018;**558**: 297-300. DOI: 10.1038/s41586-018-0186-z

[39] Robert HS, Friml J. Auxin and other signals on the move in plants. Nature Chemical Biology. 2009;**5**:325-332. DOI: 10.1038/nchembio.170

[40] Petrásek J, Friml J. Auxin transport routes in plant development. Development (Cambridge, England). 2009;**136**:2675-2688. doi:10.1242/dev.030353

[41] Holt AL, van Haperen JM, Groot EP, Laux T. Signaling in shoot and flower meristems of Arabidopsis thaliana. Current Opinion in Plant Biology. 2014;**17**:96-102. DOI: 10.1016/j.pbi.2013.11.011

[42] Su YH, Zhao XY, Liu YB, Zhang CL, O'Neill SD, Zhang XS. Auxin-induced WUS expression is essential

for embryonic stem cell renewal during somatic embryogenesis in Arabidopsis. Plant Journal for Cell and Molecular Biology. 2009;**59**:448-460. DOI: 10.1111/j.1365-313X.2009.03880.x

[43] Wu X, Dabi T, Weigel D. Requirement of homeobox gene STIMPY/WOX9 for Arabidopsis meristem growth and maintenance. Current Biology. 2005;**15**. DOI: 10.1016/j.cub.2004.12.079

[44] Elhiti M, Stasolla C. Ectopic expression of the Brassica SHOOTMERISTEMLESS attenuates the deleterious effects of the auxin transport inhibitor TIBA on somatic embryo number and morphology. Plant Science. 2011;**180**:383-390. DOI: 10.1016/j.plantsci.2010.10.014

[45] McSteen P. Auxin and monocot development. Cold Spring Harbor Perspectives in Biology. 2010;**2**:a001479. DOI: 10.1101/cshperspect.a001479

[46] Han W, Rong H, Zhang H, Wang M-H. Abscisic acid is a negative regulator of root gravitropism in Arabidopsis thaliana. Biochemical and Biophysical Research Communications. 2009;**378**:695-700. DOI: 10.1016/j.bbrc.2008.11.080

[47] Yoon E, Yang J, Lee W. Auxin and abscisic acid responses of auxin response factor 3 in Arabidopsis lateral root development. Journal of Plant Biology. 2010;**53**:150154. DOI: 10.1007/s12374-010-9100-4

[48] Rikiishi K, Matsuura T, Ikeda Y, Maekawa M. Light inhibition of shoot regeneration is regulated by endogenous abscisic acid level in calli derived from immature barley embryos. PLoS One. 2015;**10**:e0145242. DOI: 10.1371/journal.pone.0145242

[49] Fernando S, Gamage CK. Abscisic acid induced somatic embryogenesis in immature embryo explants of coconut (*Cocos nucifera L.*). Plant Science. 2000;**151**:193-198. DOI: 10.1016/s0168-9452(99)00218-6

[50] Fujii H, Zhu J-K. Osmotic stress signaling via protein kinases. Cellular and Molecular Life Sciences: CMLS. 2012;**69**:3165-3173. DOI: 10.1007/s00018-012-1087-1

[51] Fujita Y, Fujita M, Shinozaki K, Yamaguchi-Shinozaki K. ABA-mediated transcriptional regulation in response to osmotic stress in plants. Journal of Plant Research. 2011;**124**:509-525. DOI: 10.1007/s10265-011-0412-3

[52] Nakayama N, Smith RS, Mandel T, Robinson S, Kimura S, Boudaoud A, et al. Mechanical regulation of auxin-mediated growth. Current Biology. 2012;**22**:1468-1476. DOI: 10.1016/j.cub.2012.06.050

[53] Xu K, Liu J, Fan M, Xin W, Hu Y, Xu C. A genome-wide transcriptome profiling reveals the early molecular events during callus initiation in Arabidopsis multiple organs. Genomics. 2012;**100**:116-124. DOI: 10.1016/j.ygeno.2012.05.013

[54] Urano K, Kurihara Y, Seki M, Shinozaki K. "Omics" analyses of regulatory networks in plant abiotic stress responses. Current Opinion in Plant Biology. 2010;**13**:132-138. DOI: 10.1016/j.pbi.2009.12.006

[55] Martin A, Cuadrado Y, Guerra H, Gallego P, Hita O, Martin L, et al. Differences in the contents of total sugars, reducing sugars, starch and sucrose in embryogenic and non-embryogenic calli from Medicago arborea L. Plant Science. 2000;**154**:143151. DOI: 10.1016/s0168-9452(99)00251-4

[56] Huang W-L, Tsung Y-C, Liu L-F. Osmotic stress promotes shoot regeneration in immature embryo-derived callus in rice (*Oryza sativa* L.).

Journal of the Agricultural Association of China. 2002;**3**(1):76-86

[57] Huang W-L, Liu L-F. Promotion of shoot regeneration from rice (*Oryza sativa* L.) callus induced on the medium containing high concentration of sucrose. Chinese Agronomy Journal. 1998;**8**:91-100

[58] Huang W-L, Liu L-F. Changes of protein patterns during rice (*Oryza sativa* L.) callus induction and shoot regeneration induced by high concentration of sucrose. Chinese Agronomy Journal. 1999;**9**:221-231

[59] Iraqi D, Tremblay FM. Analysis of carbohydrate metabolism enzymes and cellular contents of sugars and proteins during spruce somatic embryogenesis suggests a regulatory role of exogenous sucrose in embryo development. Journal of Experimental Botany. 2001;**52**:2301-2311. DOI: 10.1093/jexbot/52.365.2301

[60] Aoki N, Whitfeld P, Hoeren F, Scofield G, Newell K, Patrick J, et al. Three sucrose transporter genes are expressed in the developing grain of hexaploid wheat. Plant Molecular Biology. 2002;**50**:453462. DOI: 10.1023/a:1019846832163

[61] Hirose T, Takano M, Terao T. Cell wall invertase in developing rice caryopsis: Molecular cloning of OsCIN1 and analysis of its expression in relation to its role in grain filling. Plant and Cell Physiology. 2002;**43**:452-459. DOI: 10.1093/pcp/pcf055

[62] Huang W-L, Wang Y-C, Lee P-D, Liu L-F. The regenerability of rice callus is closely related to starch metabolism. Taiwanese Journal of Agricultural Chemistry and Food Science. 2006;**44**:100-107. DOI: 10.6578/TJACFS.2006.012

[63] Swarup R, Kramer EM, Perry P, Knox K, Leyser OH, Haseloff J, et al. Root gravitropism requires lateral root cap and epidermal cells for transport and response to a mobile auxin signal. Nature Cell Biology. 2005;**7**:10571065. DOI: 10.1038/ncb1316

[64] Hong J, Chu H, Zhang C, Ghosh D, Gong X, Xu J. A quantitative analysis of stem cell homeostasis in the Arabidopsis columella root cap. Frontiers in Plant Science. 2015;**6**:206

[65] Brown C, Brooks F, Pearson D, Mathias RJ. Control of embryogenesis and organogenesis in immature wheat embryo callus using increased medium osmolarity and abscisic acid. Journal of Plant Physiology. 1989;**133**(6):727-733

[66] Lakshmanan P, Taji A. Somatic embryogenesis in leguminous plants. Plant Biology. 2000;**2**:136-148. DOI: 10.1055/s-2000-9159

[67] Kaur P, Kothari S. In Vitro culture of Kodo millet: Influence of 2,4-D and picloram in combination with kinetin on callus initiation and regeneration. Plant Cell, Tissue and Organ Culture. 2004;**77**:7379. DOI: 10.1023/b:ticu.0000016505.20448.44

Chapter 4

Jasmonates: An Emerging Approach in Biotic and Abiotic Stress Tolerance

Shivani Lalotra, Akhouri Hemantaranjan, Bhudeo Rana Yashu, Rupanshee Srivastava and Sandeep Kumar

Abstract

Plant hormones acts as key signaling compounds in plant stress responses and development under biotic and abiotic stresses. The potential roles of phytohormones had been considered so far and copious investigation is going on in finding the impending role of phytohormones in abiotic and biotic stresses. In the list of known classical plant hormones, Jasmonates, [jasmonic acid and its methyl ester (methyl Jasmonates)] have been recently added and shown as potential tool in enhancing tolerance of plants against various physiological processes. These are oxidized lipids (oxylipins) mainly derived from α-linolenic acids (α-LAs), play an active role in senescence through signaling, flower nectar secretion, Gprotein signaling, physiological activities and development in plants. Exogenous application of jasmonates on different plant parts have proved effective in improving plant abiotic stress tolerance particularly salinity, drought, and temperature (low/high) conditions and also in biotic stress tolerance like pathogen attack or wounding by production of defensive secondary metabolites, through the stimulation of phenyl-propanoid metabolism resulted from accumulation of JA in plant cells or tissues. Based on this prolific role of Jasmonates and its derivatives in different fields of biological sciences these phytohormones have opened new vistas and increasing interest in future Agricultural, Biotechnological and Physiological researches.

Keywords: biotic and abiotic stress, phytohormones, jasmonates and its derivatives, biosynthesis, emerging roles, signal transduction etc.

1. Introduction

The total geographical area of India is 328.7 million hectares [1], and its economy is chiefly dependent on agriculture sector. About 54.6% of the Indian population is affianced in agriculture and allied activities [2], and about 141.4 million hectares is the reported net sown area, and 200.9 million hectares is the gross cropped area with a cropping intensity of 142%. Moreover, the human population is rapidly increasing and needs a substantial increase in agricultural productivity worldwide. There has been a continuous decline in the share of agriculture and allied sector due to the changing climatic scenario, exploding population, and stressful environment which made it unfeasible for proper operation of gross cropped area. Therefore, exploiting recent and innovative strategies, tools, approaches, chemicals, and

IntechOpen

technologies is the only solution to increase productivity for the ever-increasing population. Given the importance of agriculture sector, steps have been taken to improve productivity on a sustainable basis by agronomic, biotechnological, genetical, and physiological approaches. Stressful environments are now being recognized as a potential agricultural threat for the sustainable agriculture. The commencement of environmental stresses results in plant defense responses, e.g., expression of stress-responsive genes and production of many defensive proteins and nonprotein compounds through various signaling pathways [3]. Plants produce various volatile or nonvolatile endogenous compounds and also certain mechanisms developed and deployed by plants to counteract environmental stress. Such endogenous compounds are called phytohormones.

2. Phytohormones

Phytohormones are the chemical compounds produced endogenously in very low concentrations that play significant role in the regulation and expression of gene encoding proteins. They are the diverse group of signaling molecules that result in a variety of cellular and developmental processes, signaling networks in plants under biotic and abiotic stress. They work as chemical messengers to communicate cellular activities and act either at their site of synthesis or elsewhere in plants following their transport in higher plants [4, 5]. Different phytohormones interact with each other and show synergetic or antagonist interactions that might be helpful in tolerance mechanisms. A large number of phytohormones are studied to date; among them jasmonates are the emerging players in environmental stress tolerance.

3. Jasmonate: a potent phytohormone

Jasmonates [jasmonic acid (**Figure 2**) and its methyl ester methyl jasmonate (**Figure 1**)] the cyclopentanone phytohormones are a class of oxidized lipids (oxylipins) derived from α-linolenic acids (α-LAs) through lipoxygenase-dependent manner. [6] first isolated JA from culture filtrate of the fungus *Lasiodiplodia (Botryodiplodia) theobromae,* a plant pathogen, was identified as a plant growth inhibitor, whereas its derivative methyl jasmonate was first isolated from *Jasminum grandiflorum* (jasmine) petal extract [7]. (+)-7-iso-Jasmonoyl-L-isoleucine (JA-Ile) is the best-described bioactive JA [7] till date, but other JAs like *cis*-jasmone, jasmonoyl ACC (JA-ACC), and jasmonoyl isoleucine (JA-Ile) are also studied by scientists with multiple biological functions [9–12]. Various developmental and environmental factors are responsible for the production of JA in membranes and resulted in expression of stress tolerant genes (**Figure 3**). JA is ubiquitously found in the plant kingdom and results in the expression of genes at the transcriptional and post transcriptional levels [13, 14]. An imperative role of jasmonate (when applied exogenously in low concentrations) is reported in enhanced pathogen resistance.

Cross talks of jasmonates with signaling pathways generated by various phytohormones like ABA, salicylic acid, ethylene, etc. result in diverse developmental processes like seed germination, seedling growth, pollen fertility, fruit ripening, senescence and tolerance. However, its extent of effectiveness entirely depends on the type of plant species tested or its concentration. MeJA is more volatile than JA, so exposure to it either in solution or in the gaseous phase can elicit plant responses. Apart from its significant role in plants, derivatives of jasmonates, e.g., methyl jasmonate are used as a fragrant constituent in many aromatic mixtures [15].

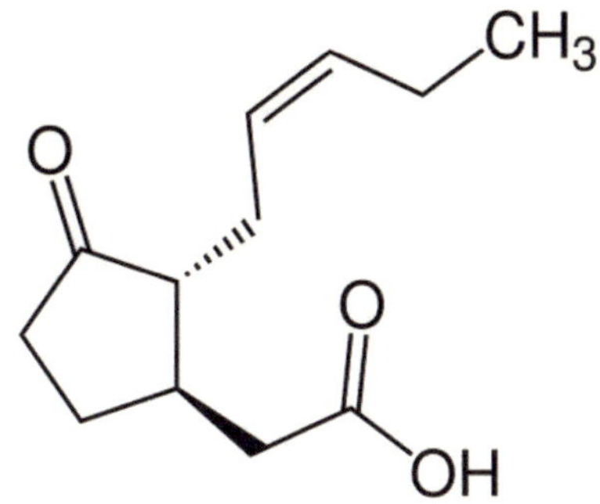

Figure 1.
Chemical structure of methyl jasmonate [3-oxo-2-(2-pentenyl)-, methyl ester].

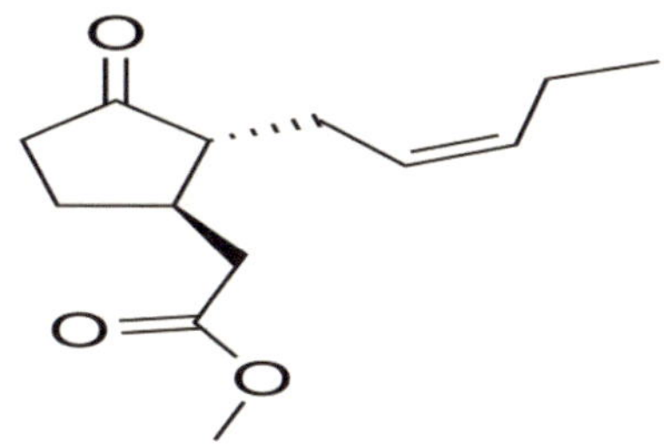

Figure 2.
Chemical structure of Jasmonic acid [3-oxo-2-(2-pentenyl) cyclopentaneacetic acid].

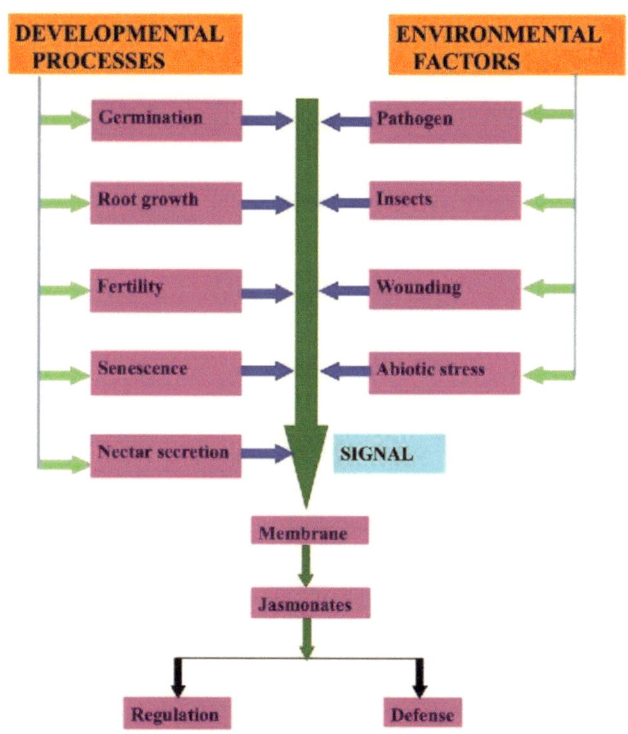

Figure 3.
Various developmental and environmental factors are shown that signals membrane for the production of endogenous jasmonates which further results in the regulation of certain genes and defense mechanisms against stressful environment. Source: Ahmad et al., [16].

4. Biosynthesis of jasmonates: the octadecanoid pathway

The biosynthetic pathway of jasmonates (**Figure 4**) was identified in Year 1984. The scientists Vick and Zimmerman [17] were the first to illustrate the biosynthetic pathway in a simplified manner which indicated that linolenic acid could be converted into the cyclopentanone 12-oxo-phytodienoic acid (12-oxo-PDA) through lipoxygenase enzyme. Methodically the pathway was studied in model plants like Arabidopsis and tomato. The octadecanoid pathway of Jasmonates completes in two cellular organelles such as chloroplasts and peroxisomes and are considered to be the primary sites [7, 18]. α-Linolenic acid (α-LeA) released from galactolipids (due to wounding or pathogens attack) of chloroplast membranes is found to be the main player of MeJA and JA production [12]. Phospholipases1 (PL1), lipoxygenase, allene oxide synthase, and allene oxide cyclase (AOC) are significant enzymes involved in biosynthesis of jasmonates.

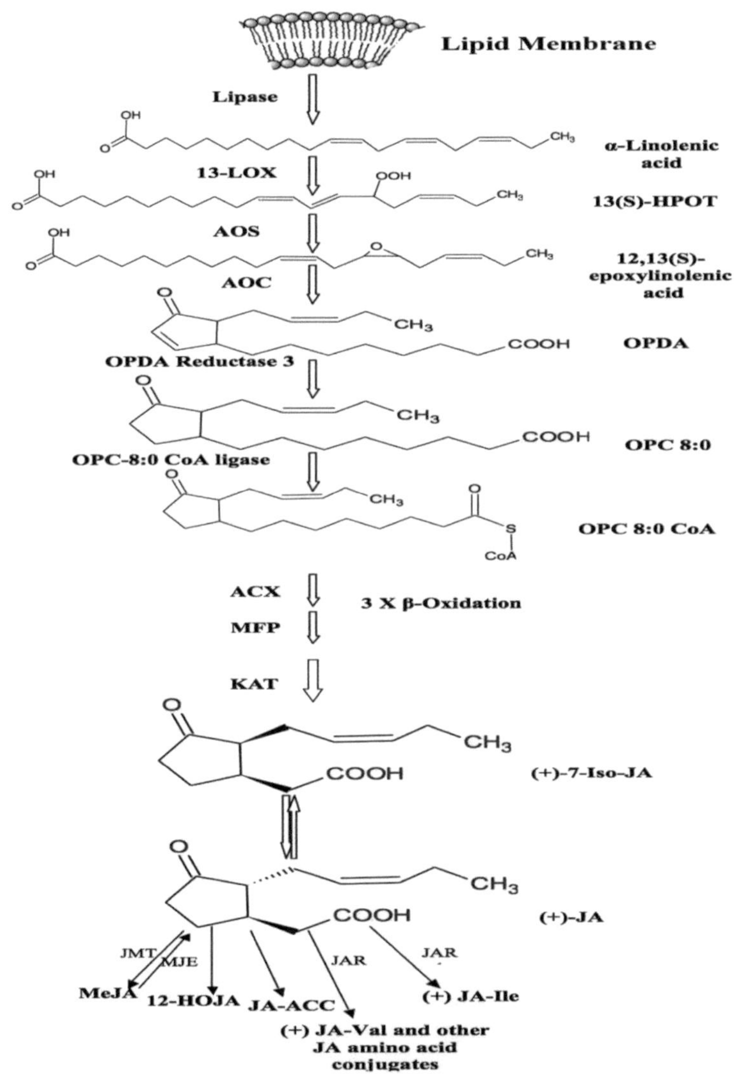

Figure 4.
Step wise pathway of jasmonates biosynthesis. Source: Ahmad et al., [16].

4.1 Role of enzymes involved in jasmonates biosynthetic pathway

- **Phospholipases:** formation of α-LeA from chloroplastic membrane lipids
- **13-Lipoxygenase (LOX):** addition of oxygen molecule to α-LeA and results in formation of an intermediate compound 13-hydroperoxy-9,11,15-octadecatrienoic acid (13-HPOT).
- **Allene oxide synthase (AOS):** oxidation of 13-HPOT to allene oxide
- **Allene oxide cyclase (AOC):** formation of 12-oxo-phytodienoic acid (12-OPDA) an unstable compound from allene oxide.

The AOS and AOC are present in plastids and they act in concert [19]. 12-OPDA is the final product of biosynthetic pathway formed in chloroplast and undergoes three cycles of β-oxidation in the peroxisomes [8] A methylation reaction by JA methyl transferase results in the formation of jasmonate methyl ester derivative, i.e., methyl jasmonate. Among the six 13-LOXs of Arabidopsis, four of them are (LOX2, LOX3, LOX4, and LOX6) but LOX2 is a vital lipoxygenase in JA biosynthesis.

4.2 Key roles of LOX2

- Responsible for the bulk of JA formation upon wounding [7, 20]
- Oxylipin generation during natural and dark-induced senescence

5. Jasmonates signal perception and transduction pathway

Several transcription factors (TFs), repressors, up- regulation and down-regulation of certain genes, and members of ubiquitin-proteasome complexes are involved in JA signal perception and transduction pathway (**Figure 5**). Different independent studies [21, 22] with GC-MS and HPLC analyses in *A. thaliana* revealed that (+)-7-iso-JA-L-Ile is the only natural and direct JA-signaling ligand in plants.

The Skp1/Cullin/F-box (SCF) complex is of proteinous type and is the ubiquitin-proteasome complex. In jasmonate signal transduction pathway *coronatine insensitive1* (*COI1*), the locus encodes an F-box protein that associates with its other counterparts, SKP1, Cullin, and Rbx proteins, to form an E3 ubiquitin ligase [23]. So *COI1* is found to be the jasmonate receptor in signaling pathway. In 2007, [22, 24, 25] scientists in independent research found new family of proteins called jasmonate zim-domain (JAZ) proteins in Arabidopsis, which were found to be the key negative regulator of JA signaling and act as substrate for SCF^{COI1} E3 ubiquitin ligase complex. About 12 JAZ proteins were present in *A. thaliana* and contain N-terminal domain, a highly conserved C-terminal Jas domain that mediates the interaction with the *COI1* and several transcription factors, and the conserved protein–protein interaction domain, the ZIM (TIFY) domain that helps in JAZ dimerization and interaction with novel interactor of JAZ (NINJA).

Moreover, the interaction of *COI1* with the Jas domain of JAZ proteins in the presence of JA-Ile forms the co-receptor complex [26] SCF^{COI1}, E3 ubiquitin ligase complex, JAZ degrons (*JAZ1* to *JAZ12*) and IP5 form the co-receptor complex and found to be true jasmonates receptors [27].

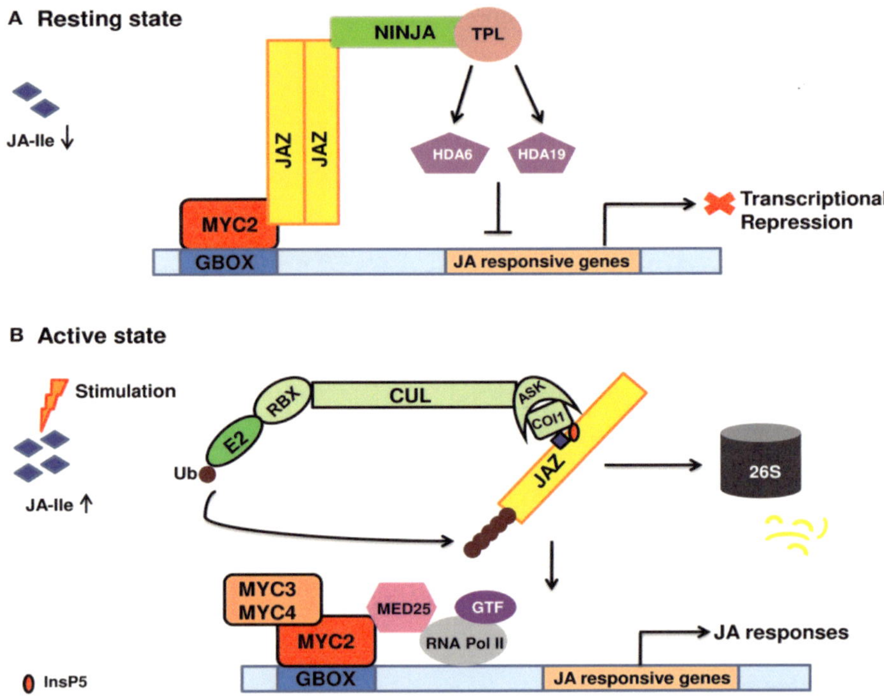

Figure 5.
Signal transduction pathway of jasmonic acid.

A number of co-repressors also play their role simultaneously in signal transduction pathway. Corepressors in transcriptional regulators inhibit transcription initiation. TOPLESS (TPL) and TPL-related proteins (TPRs) are members of Groucho/Tup1 co-repressor family that causes histone deacetylation and demethylation which resulted in chromatin modification, eventually mediated repression of genes. [10]. Interaction of TPL with JAZ proteins takes place via ethylene response factor (ERF)-associated amphiphilic repression (EAR) motif. Those JAZ proteins having absence of repression motif recruit TPL through an adapter protein called novel interactor of JAZ (NINJA) [28].

MYC2 the master regulator of many biological processes is a bHLH transcription factor mediating the transcriptional regulation of JA. Other transcription factors that controls JA mediated signaling are MYC3, MYC4, MYB, GL3, EGL3 AP, GL1, etc.

5.1 Role of MYC2 in JA mediating signal transduction

- An activator of JA-induced root growth inhibition
- Anthocyanin biosynthesis, oxidative stress tolerance
- Mediating resistance to necrotrophic pathogens, insects
- Biosynthesis of tryptophan and indole glucosinolates

Various homologs of MYC2 like MYC3 and MYC4 binds with G-box (5′-CACGTG-3′) and G-box-related hexamers and regulate the transcription of downstream targets [29].

6. Emerging roles of Jasmonates

6.1 Inhibitory action on seedling growth

Exogenous application of JA possibly shows inhibition of primary root growth, leaf expansion, and hypocotyl elongation which ultimately leads to inhibitory action in seedling growth [30]. InsP5 enhances the interaction of COI1 with JAZ9 and the inhibitory effect of JAs on root growth [31]. JA represses leaf expansion by inhibiting the activity of the mitotic cyclin CycB1;2 and cell division, rather than by affecting cell size. Transcription factor like MYC2 and its close homologs shows both positive and negative effects on hypocotyls in red/far-red light and blue light conditions; it works positively in inhibition of hypocotyl elongation in red/far-red light and negatively regulates the inhibition of hypocotyl elongation by blue light. ERF109 binds to and activates *anthranilate synthase A1* (*ASA1*) and *YUCCA2* (promoters of the auxin biosynthetic genes) and results in the promotion of lateral root formation in Arabidopsis [32].

6.2 Role in plant reproductive development

Various transcription factors of R2R3-MYB family like MYB21, MYB24, and MYB57 are direct targets of JAZ proteins (**Figure 6**). These TFs have significant role in mediating JA-regulated stamen development [33]. Formation of MYB-MYC complexes due to the association of MYB21 and MYB24 with the IIIe bHLH factors MYC2, MYC3, MYC4, and MYC5 controls stamen development in Arabidopsis [34]. Besides the role of JA on stamen development, JA plays a major role in seed and embryo development in tomato The *jasmonic acid-insensitive1* (*jai1*) mutant, which exhibits a loss of function of the tomato homolog of COI1, cannot set viable seeds. Moreover, production of OPDA and a residual amount of JA, in tomato mutant *acx1a*, set viable seeds. Gene silencing (*OPR3* silenced gene) in *SiOPR3* a transgenic line of tomato produces comparable amount of OPDA to wild type and sets only a few viable seeds; methyl-JA treatment can restore the seed setting of *SiOPR3.* This further suggests the role of methyl jasmonate in maternal control of seed development [35].

6.3 Role in abiotic stress tolerance

Numerous morphological, physiological, biochemical and molecular changes take place due to abiotic stresses like drought stress, salinity stress, high- and low-temperature stress, heavy metal toxicity, etc.; these stresses adversely affect plant growth and productivity. JA is believed to play a role in plant responses to abiotic stresses including drought, salt, and heat stress. Salinity is one of the perilous stresses that causes physiological drought and is responsible for delayed seed germination, seedling establishment and reduced growth and yield of any crop. Under salt stress, jasmonates proved to be an imperative phytohormone in mitigation. Jasmonates recovered salt inhibition on dry mass production in rice [36] and diminished the inhibitory effect of NaCl on the rate of $^{14}CO_2$ fixation, protein content in *Pisum sativum* [37]. The pleiotropic effects of MeJA in protecting plants have been reported for several plants [38], reported in his studies JA is responsible for the amelioration of chilling injury, water stress, and salinity stress in *Oryza sativa* L., *Lycopersicon esculentum* L. [39], *Fragaria vesca* [40], and *Hordeum vulgare.* High-temperature stress destructively influences plant processes and disturbs the cell homeostasis [41]. Heat shock proteins (HSPs) are synthesized in plants in response to high temperature that prevent denaturation and assist refolding of

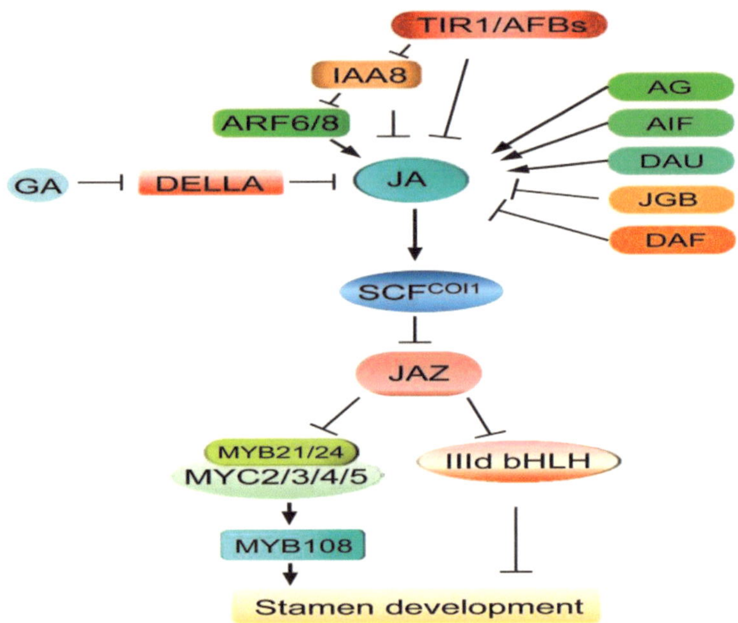

Figure 6.
JA signaling and crosstalk in stamen development (Source: Huang et al, [42]).

damaged proteins. Electrolyte leakage assays in heat-stressed plants after application of low-concentration MeJA demonstrates the cell viability responses. Heating WT Arabidopsis led to the accumulation of several jasmonates including OPDA, MeJA, JA, and JA-Ile, and the expression of jasmonate inducible gene PDF1.2 was found to be high upon heat stress exposure. Suppressor of G2 allele of SKP1 (SGT1) protein operates as a cofactor of heat shock protein 90 (HSP90) in both plants and mammals forming functional complexes and providing thermotolerance. JA also showed essential role in heavy metal and nutrient toxicity. In a study by [43], they reported that the excess amounts of boron present in soil decrease the net photosynthetic rate, closing of stomata, internal CO_2 concentration, and total chlorophyll content in leaves. Foliar application of boron-stressed plants of Artemisia with MeJA started to stimulate the synthesis of antioxidant enzymes, reduce the amount of lipid peroxidation, and enhance artemisinin content. [44] found that the first report on jasmonate-induced anticancer activities exhibited their capacity to cause both cell death and suppression of cell proliferation. MJ was studied in topical application for precancerous and cancerous skin lesions [45].

6.4 Role in biotic stress tolerance

Two types of responses are shown by plants due to tissue injury, and jasmonates plays a significant role in these responses by signaling in plants. Two types of responses are local response and systemic response (**Figure** 7).

In local response during tissue damage, various attacker-derived signals or damaged-associated plant-derived signals are produced; these signals are either chemical or physical in nature and are recognized by PRRs pattern recognition receptors present on cell surface. This recognition event activates *de novo* synthesis of JA and JA-Ile by an unknown pathway. SCF^{COI1}/26 proteasome activation by JA-Ile results in degradation of JAZ proteins. These proteins are responsible for the repression of transcription factors (TFs) involved in the expression of defense-related traits.

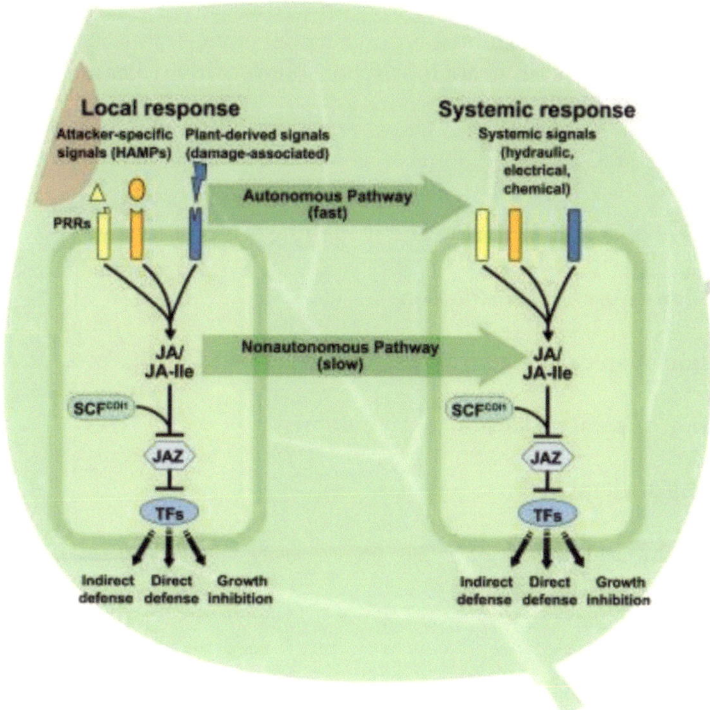

Figure 7.
Role of JA in tissue damage (source: Abraham et al. [46]).

Systemic responses are mediated by two distinct pathways involving JA. Cell-nonautonomous pathway is a slow pathway and cell autonomous pathway is a fast pathway in defense responses. In cell-nonautonomous pathway upon leaf damage, JA is produced and translocated to undamaged leaf where it triggers JA responses in target cells whereas, in cell autonomous pathway, wound-induced production of a mobile signal (other than JA) activates JA/JA-Ile synthesis and subsequent responses in distal tissues. To optimize the spatial and temporal expression of responses the two pathways may work synergistically.

6.5 Role in the promotion of leaf senescence and seed germination

Exogenous application of low concentration of methyl jasmonate below 1 micro molar do not promote premature leaf senescence, but if the concentration increases beyond 30 micro molar, symptoms of premature leaf senescence were seen in the early stages of plant growth by upregulating the expression of senescence-associated genes and by downregulating photosynthesis-related genes. JA promotes leaf senescence in a COI1-dependent manner; TFs MYC2, MYC3, and MYC4 mediate JA/dark-induced leaf senescence by upregulating the expression of senescence-associated genes (e.g. *senescence-associated gene 29* [*SAG29*]) and chlorophyll catabolic enzyme genes (CCGs) (e.g. *pheophorbide A oxygenase*), as well as by down-regulating photosynthesis-related genes (e.g., *chlorophyll A/B binding protein 1*), whereas in germination process, JA delays the ABA-mediated inhibition of seed germination in Arabidopsis. During the cold-stimulated germination of wheat (*Triticum aestivum*) seeds, JA biosynthesis-related gene expression and JA biosynthesis increase rapidly in the dormant embryos after transfer to room temperature, and JA suppresses ABA biosynthesis to promote cold-stimulated germination [47].

Other key roles of jasmonates are:

As most of the work of jasmonates did in model plant Arabidopsis, so numerous roles of JA could be found in Arabidopsis, but some positive roles are also found in other plants.

- Delay of flowering in Arabidopsis
- Regulation of stomatal closure and reopening in Arabidopsis
- Promotion of trichome formation
- Inhibition of apical hook formation in Arabidopsis
- Inhibition of petal expansion in Arabidopsis
- Gravitropism in plants
- Sex determination in maize

7. Conclusion

The twin challenges like declining food and nutritional security and changing climatic scenario require the use of such biotechnological, physiological, genetical, and agronomical strategies to accomplish the demands of food to exploding population. Increasing abiotic stress made land barren or unutilizable for crop production. Tolerance to biotic and abiotic stresses is a challenge of agro-economic impact. Moreover, in a field condition additive effect of a number of stresses may further result in decreased soil fertility and crop yields. Researches on tolerance mechanisms and cross talks by phytohormones proved to be constructive for researchers in developing tolerant plants against particular stress. From the above information on jasmonates, the role of jasmonates in plant development is very well established but there is limited information available in the literature on how plant processes vary from species to species with JA application; a large number of components/genes of JA/MeJA including its receptors have been identified, but their appropriate functions still need to be explored, also various genes involved in growth regulation at different stages of development are yet to be identified.

DOI: http://dx.doi.org/10.5772/intechopen.84608

Author details

Shivani Lalotra[1*], Akhouri Hemantaranjan[1], Bhudeo Rana Yashu[1], Rupanshee Srivastava[2] and Sandeep Kumar[3]

1 Department of Plant Physiology, Institute of Agricultural Sciences, Banaras Hindu University, Varanasi, India

2 Department of Genetics and Plant Breeding, Institute of Agricultural Sciences, Banaras Hindu University, Varanasi, India

3 Department of Agronomy, Institute of Agricultural Sciences, Banaras Hindu University, Varanasi, India

*Address all correspondence to: shivani90lalotra@gmail.com

References

[1] Central Statistics Office. Ministry of Statistics and Programme Implementation, Govt. of India : Land use statistics. pp. 2014-2015

[2] Annual report 2016-17, Department of Agriculture, Cooperation & Farmers Welfare: census 2011

[3] Ghasemzadeh A, Jaafar HZE. Interactive effect of salicylic acid on some physiological features and antioxidant enzymes activity in ginger (Zingiber officinale roscoe). *Molecules*. 2013;**18**:5965-5979

[4] Vob U, Bishopp A, Farcot E, Bennett MJ. Modelling hormonal response and development. *Trends in Plant Science*. 2014;**19**:311-319

[5] Peleg Z, Blumwald E. Hormone balance and abiotic stress tolerance in crop plants. *Current Opinion in Plant Biology*. 2011;**14**(2011):290-295

[6] Aldridge DC, Galt S, Giles D, Turner WB. Metabolites of *Lasiodiplodia theobromae*. *Journal of the Chemical Society C*. 1971;**627**:1623-1621

[7] Sembdner G, Parthier B. The biochemistry and physiology and molecular actions of jasmonates. *Annual Review of Plant Physiology and Plant Molecular Biology*. 1993;**44**:569-586

[8] Delker C, Stenzel I, Hause B, Miersch O, Feussner I, Wasternack C. Jasmonate biosynthesis in *Arabidopsis thaliana* - enzymes, products, regulation. *Plant Biology*. 2006;**8**:297-306. DOI: 10.1055/s-2006-923935

[9] Avanci NC, Luche DD, Goldman GH, Goldman MHS. Jasmonates are phytohormones with multiple functions. including plant defense and reproduction. *Genetics and Molecular Research*. 2010;**9**:484-505. DOI: 10.4238/vol9- 1gmr754

[10] Macrae R, Long JA. Interactions between TOPLESS and histonemodifying enzymes. In: 22nd International Conference on Arabidopsis Research. Madison, WI: The North American Arabidopsis Steering Committee. 2011

[11] Rohwer C. L, Erwin J. E : Horticultural applications of jasmonates : a review. *The Journal of Horticultural Science and Biotechnology*. 2008;**83**:283-304. DOI: 10.1016/j.biotechadv.2013.09.009

[12] Wasternack C, Kombrink E. Jasmonates: Structural requirements for lipid- derived signals active in plant stress responses and development. ACS Chemical Biology. 2010;**5**:63-77

[13] Pirbalouti A. G, Mirbagheri H, Hamedi B, Rahimi E: Antibacterial activity of the essential oils of myrtle leaves against *Erysipelothrix rhusiopathiae*. *Asian Pacific Journal of Tropical Biomedicine*. 2014;**4**:505-509. DOI: 10.12980/APJTB.4.2014B1168

[14] Schaller M. The behavioral immune system and the psychology of human sociality. *Philosophical Transactions of the Royal Society B*. 2011;**366**:3418-3426. DOI: 10.1098/rstb.2011.0029

[15] Staswick PE. Jasmonate, genes, and fragrant signals. *Plant Physiology*. 1992;**99**:804-807

[16] Ahmad P, Rasool S, Gul A, Sheikh S, Akram AN, Ashraf M, et al. Jasmonates: Multifunctional roles in stress tolerance. *Frontiers in Plant Science*. 2016;**7**:813. DOI: 10.3389/fpls.2016.00813

[17] Vick BA, Zimmerman DC. Biosynthesis of jasmonic acid by several plant species. Plant Physiology. 1984;**75**:458-461. DOI: 10.1104 75.2.458

[18] Cheong JJ, Choi YD. Methyl jasmonate as a vital substance

in plants. *TRENDS in Genetics*. 2003;**19**:409-413

[19] Schaller A, Stintzi A. Enzymes in jasmonate biosynthesis- structure, function, regulation. *Phytochemistry*. 2009;**70**:1532-1538. DOI: 10.1016/j.phytochem.2009.07.032

[20] Glauser G, Dubugnon L, Mousavi SAR, Rudaz S, Wolfender J-L, Farmer EE. Velocity estimates for signal propagation leading to systemic jasmonic acid accumulation in wounded Arabidopsis. *Journal of Plant Growth Regulation*. 2009;**284**:34506-34513

[21] Fonseca S, Chini A, Hamberg M, Adie B, Porzel A, Kramell R. (+)-7-iso- Jasmonoyl-L-isoleucine is the endogenous bioactive jasmonate. *Nature Chemical Biology*. 2009a;**5**:344-350

[22] Thines B, Katsir L, Melotto M, Niu Y, Mandaokar A, Liu G. JAZ repressor proteins are targets of the SCF (COI1) complex during jasmonate signalling. *Nature*. 2007;**448**:661-665. DOI: 10.1038/nature05960

[23] Xie DX, Feys BF, James S, Nieto-Rostro M, Turner JG. COI1:an Arabidopsis gene required for jasmonate-regulated defense and fertility. *Science*. 1998;**280**:1091-1094. DOI: 10.1126/science.280.53 66.1091

[24] Chini A, Fonseca S, Fernández G, Adie B, Chico JM, Lorenzo O. The JAZ family of repressors is the missing link in jasmonate signalling. *Nature*. 2007;**448**:666-671. DOI: 10.1038/nature06006

[25] Yan Y, Stolz S, Chételat A, Reymond P, Pagni M, Dubugnon L. A down stream mediator in the growth repression imb of the jasmonate pathway. Plant Cell. 2007;**19**:2470-2483. DOI: 10.1105/tpc.107. 050708

[26] Yan J, Zhang C, Gu M, Bai Z, Zhang W. Qi: The Arabidopsis CORONATINE INSENSITIVE1 protein is a jasmonate receptor. *PlantCell*. 2009;**21**:2220-2236. DOI: 10.1105/tpc.109.065730

[27] Sheard LB, Tan X, Mao H, Withers J, Ben-Nissan G, Hinds TR. Jasmonate perception by inositol-phosphate-potentiated COI1-JAZ co-receptor. *Nature*. 2010;**468**:400-405. DOI: 10.1038/nature09430

[28] Pauwels L, Barbero GF, Geerinck J, Tilleman S, Grunewald W, Pérez AC. NINJA connects the co-repressor TOPLESS to jasmonate signalling. *Nature*. 2010;**464**:788-791. DOI: 10.1038/nature08854

[29] Fernández-Calvo P, Chini A, Fernández-Barbero G, Chico JM, Gimenez- Ibanez S, Geerinck J. The Arabidopsis bHLH transcription factors MYC3 and MYC4 are targets of JAZ repressors and act additively with MYC2 in the activation of jasmonate responses. *Plant Cell*. 2011;**23**:701-715. DOI: 10.1105/tpc.110.080788

[30] Wasternack C, Hause B. Jasmonates: biosynthesis, perception, signal transduction and action in plant stress response, growth and development. *Annals of Botany*. 2013;**111**:1021-1068

[31] Mosblech A, Thurow C, Gatz C, Feussner I, Heilmann I. Jasmonic acid perception by COI1 involves inositol polyphosphates in *Arabidopsis thaliana*. *Plant Journal*. 2011;**65**:949-957

[32] Sun J, Chen Q, Qi L. Jasmonate modulates endocytosis and plasma membrane accumulation of the Arabidopsis PIN2 protein. The New Phytologist. 2011;**191**:360-375

[33] Cheng H, Song S, Xiao L, Soo HM, Cheng Z, Xie D, et al. Gibberellin acts through jasmonate to control the expression of *MYB21*, *MYB24*, and *MYB57* to promote stamen filament growth in *Arabidopsis*. *PLoS Genetics*. 2009;**5**:e1000440

[34] Qi T, Huang H, Song S, Xie D. Regulation of jasmonate-mediated stamen development and seed production by a bHLH-MYB complex in Arabidopsis. *The Plant Cell*. 2015;**27**:1620-1633

[35] Scalschi L, Sanmartin M, Camanes G, Troncho P, Sanchez-Serrano JJ, Garcia- Agustin P, et al. Silencing of *OPR3* in tomato reveals the role of OPDA in callose deposition during the activation of defense responses against *Botrytis cinerea*. *The Plant Journal*. 2015;**81**:304-315

[36] Kang JH, Wang L, Giri A, Baldwin IT. Silencing threonine deaminase and JAR4 in *Nicotiana attenuata* impairs jasmonic acid–isoleucine- mediated defenses against *Manduca sexta*. *Plant Cell*. 2006;**18**:3303-3320. DOI: 10.1105/tpc.106.041103

[37] Velitcukova M, Fedina I. Response of photosynthesis of *Pisum sativum* to salt stress as affected by methyl jasmonate. *Photosynthetica*. 1998;**35**:89-97

[38] Lee IJ, Foster KR, Morgan PW. Photoperiod control of gibberellin levels and flowering in Sorghum. *Plant Physiology*. 1998;**116**:1003-1010

[39] Ding Y, Kalo P, Yendrek C, Sun J, Liang Y, Marsh JF. Abscisic acid coordinates nod factor and cytokinin signaling during the regulation of nodulation in *Medicago truncatula*. *Plant Cell*. 2008;**20**:2681-2695. DOI: 10.1105/tpc.108.061739

[40] Wang CY. Methyl jasmonate reduces water stress in strawberry. *Journal of Plant Growth Regulation* 1999;**18**: 127-134. DOI: 10.1007/PL00007060

[41] Bokszczanin KL. Solanaceae Pollen Thermotolerance Initial Training Network (SPOTITN) Consortium, and Fragkoste fanakis, S. Perspectives on deciphering mechanisms underlying plant heat stress response and thermotolerance. *Frontiers in Plant Science*. 2013;**4**:315. DOI: 10.3389/fpls.2013.00315

[42] Huang H, Liu B, Liu L, Song S: Jasmonate action in plant growth and development. *Journal of Experimental Botany*. 2016;**68**(6):1349-1359

[43] Aftab T, Khan MM. A, Idrees M, Naeem M, Hashmi: Methyl jasmonate counteracts boron toxicity by preventing oxidative stress and regulating antioxidant enzyme activities and artemisin in biosynthesis in *Artemisia annua* L. Protoplasma 2011;**248**:601-612

[44] Halitschke R, Baldwin IT. Jasmonates and related compounds in plant insect interactions. *J. Plant Growth Regul*. 2004;**23**:238-245

[45] Pedranzani H, Racagni G, Alemano S, Miersch O, Ramirez I, Pena-Cortes H, et al. 2003. Salt tolerant tomato plants show increased levels of jasmonic acid. *Plant Growth Regulation*. 2003;**41**:149-158

[46] Koo AJK, Howe GA. The wound hormone jasmonate. *Phytochemistry*. 2009;**70**(13-14):1571-1580

[47] Xu Q, Truong TT, Barrero JM, Jacobsen JV, Hocart CH, Gubler F. 2016. A role for jasmonates in the release of dormancy by cold stratification in wheat. Journal of Experimental Botany. 2016;**67**:3497-3508

Chapter 5

A Regulatory Circuit Integrating Stress-Induced with Natural Leaf Senescence

Otto Teixeira Fraga, Bruno Paes de Melo,
Luiz Fernando de Camargos, Debora Pellanda Fagundes,
Celio Cabral Oliveira, Eduardo Bassi Simoni,
Pedro Augusto Braga dos Reis
and Elizabeth Pacheco Batista Fontes

Abstract

Any condition that disrupts the ER homeostasis activates a cytoprotective signaling cascade, designated as the unfolded protein response (UPR), which is transduced in plant cells by a bipartite signaling module. Activation of IRE1/bZIP60 and bZIP28/bZIP17, which represent the bipartite signaling arms and serve as ER stress sensors and transducers, results in the upregulation of ER protein processing machinery-related genes to recover from stress. However, if the ER stress persists and the cell is unable to restore ER homeostasis, programmed cell death signaling pathways are activated for survival. Here, we describe an ER stress-induced plant-specific cell death program, which is a shared response to multiple stress signals. This signaling pathway was first identified through genome-wide expression profile of differentially expressed genes in response to combined ER stress and osmotic stress. Among them, the development and cell death domain-containing N-rich proteins (DCD/NRPs), *NRP-A* and *NRP-B*, and the transcriptional factor *GmNAC81* were selected as mediators of cell death in plants. These genes were used as targets to identify additional components of the cell death pathway, which is described here as a regulatory circuit that integrates a stress-induced cell death program with leaf senescence via the NRP-A/NRP-B/GmNAC81:GmNAC30/VPE signaling module.

Keywords: senescence, stress, NRP, DCD, BiP, NAC, VPE, ER, osmotic stress, drought

1. Introduction

The onset of leaf senescence is a highly regulated developmental program that is controlled by both genetics and the environment. Multiple stresses in plants induce programmed cell death, and the underlying regulatory mechanisms are often associated with molecular links of developmentally programmed senescence. The transcriptome changes induced by different environmental stressors are not

IntechOpen

entirely overlapping, but functional analysis of genes commonly induced as shared responses can give clues on signaling integration. This approach has been used to select for overlapping genes as candidate regulatory components that integrate the ER stress and osmotic stress responses, which were shown later to participate also in natural leaf senescence. Among genes identified as components of the ER and osmotic stress shared response, the developmental and cell death (DCD) domain-containing asparagine-rich proteins (NRP-A and NRP-B) were the first ones to be characterized as cell death-promoting proteins, and hence this multiple stress-integrating signaling was designated as stress-induced DCD/NRP-mediated cell death response. Further characterization of the cell death pathway implicated in the discovery of the signaling module ERD15/NRPs/GmNAC81:GmNAC30/VPE that also has been shown to operate in developmentally programmed leaf senescence. This plant-specific cell death signaling module, which operates in both stress-induced and natural leaf senescence, constitutes the primary focus of this chapter.

2. Modest overlapping of ER stress and osmotic stress response identifies NRPs and NACs as cell death-promoting genes

2.1 Osmotic stress responses

Organisms, in general, are continually adapting to internal and external stimuli, which activate sensor proteins to subsequently transmit the signal to downstream effectors responsible for the assembly of adaptive cellular responses [1]. Abiotic stresses consist of a set of adverse environmental conditions that limits plant development. Cold, high temperature, salinity, water availability (drought or overflow), radiation, pollution, and chemical exposure are the most common examples of types of abiotic stresses [2].

Generally, a signaling sensor network connects internal and external stimuli to adaptive responses leading to molecular modifications that allow physiological adjustments, which ultimately cause susceptibility or tolerance to the exposed conditions. Molecular responses to abiotic stress conditions in plants are crucial for survival and productivity as these stresses often limit yield. Among abiotic stresses, drought and excess salinity conditions induce sophisticated adaptive responses in plants to cope with or acclimate to these adverse environmental conditions [3, 4]. Some types of abiotic stress responses are better understood than others. In plants, for example, the molecular mechanisms of perception and responses to drought, high salinity, and endoplasmic reticulum stress are well characterized, and many stress-related cell signaling pathways are completely elucidated, revealing some convergence points between them.

The osmotic stress in plants, caused by water deprivation or high salinity, for example, undergoes a set of characteristic morphological, molecular, and physiological changes. One of the most notorious symptoms in plants under low water availability is the ABA-mediated stomatal closure [5]. This hormone-mediated morphological change affects plant physiology. The stomatal closure prevents the evapotranspiration, optimizing the cell water use, but it also compromises carbon dioxide uptake, causing imbalances on photosynthetic apparatus, which culminates on reactive oxygen species (ROS) production [6, 7]. The ROS accumulation acts as a signal to the cell, which triggers mechanisms of ROS-associated detoxification, including upregulation of antioxidant enzymes, osmolyte, and electron-carrier synthesis [8]. There is evidence that osmotic stress and temperature changes are capable of generating lipid-derived signal transducers, including the phosphatidic acid, phosphoinositides, sphingolipids, lysophospholipids, oxylipins,

N-acylethanolamines, and others. Water deprivation causes a collapse on the organization of membrane lipids, disrupting its permeability and some significant molecular interactions between lipids and proteins, which act as a cell signal to stress-mediated physiological changes. The mechanisms of how stress responses are connected with membrane lipid transducer generation are still unclear, but lipid messengers can alter protein and enzymatic functions [9].

2.2 ER stress responses

The endoplasmic reticulum is one of the most dynamics organelles in cell machinery. It is the gateway for the synthesis of secretory proteins and contains the necessary apparatus to ensure quality protein synthesis, protein maturation, and secretion in eukaryotic cells [10]. Furthermore, the ER can modulate some chronic stress-related pathways, promoting oxidative stress, autophagy, and apoptotic cell death in mammals and plant cells [11–13].

Several adverse environmental conditions can affect the ER quality control machinery, causing unfolded/misfolded protein accumulation in the ER lumen. The secretory proteins are synthesized in ER membrane-bound polysomes, and, as soon as they enter the organelle, they are processed by the ER processing machinery. Under normal conditions, there is a perfect balance between the rate of protein synthesis and ER processing capacity. Any conditions that disrupt this balance promote unfolded/misfolded protein accumulation in the ER lumen. As a consequence, the perturbation on ER function triggers a sophisticated and coordinated signal cascade, perceived by ER membrane-associated sensors, which activate the expression of ER-resident chaperones, foldases, and components of the ER quality control machinery. Collectively, these cytoprotective mechanisms are known as the unfolded protein response pathway (UPR, **Figure 1**) [14].

The detection of ER stress is mediated by membrane-associated sensors, identified both in mammals and plants. In mammals, there are three of these sensors: kinase/endoribonuclease inositol-requiring enzyme 1 (IRE1), activating transcription factor 6 (ATF6), and protein kinase RNA-like ER kinase (PERK) [15], which are regulated by the ER-resident molecular chaperone BiP (binding protein). The ER sensors initiate the UPR to restore ER homeostasis under stress condition. If the adverse physiological status is prolonged, they can initiate some alternative routes leading to cell death.

Under normal conditions, BiP is bound to the luminal domain of these receptors, keeping them inactive. With the stress progression and consequent misfolded protein accumulation, the BiP molecular chaperone function is required to prevent aggregation of the unfolded proteins. Therefore, under these stress conditions, BiP is released from the ER receptors, which leads to their activation. The three ER signal transducers act in different ways, but in convergent stress-responsive pathways. IRE1 (IRE1a and IRE1b) displays a dual biochemical activity. It harbors a ribonuclease and kinase activity at the C-terminus, responsible for the unconventional spliceosome-independent splicing of X-box binding protein 1 (XBP1) mRNA. Stress-mediated BiP release from the IRE1 N-terminus promotes IRE1 homodimerization, which sequentially activates its kinase via autophosphorylation and endoribonuclease activity, culminating on spliceosome-independent splicing of XBP1, a bZIP transcriptional factor. Under normal conditions, the XBP1u (unspliced form) is constitutively translated into a low-functional transcription factor, which is rapidly degraded by the proteasome and does not effectively activate UPR. The IRE1-mediated mRNA splicing removes an unconventional intron of 26 nucleotides, which causes a shifting frame in XBP1 mRNA translation, generating a protein of 376 amino acids instead of 261 amino acids when unprocessed.

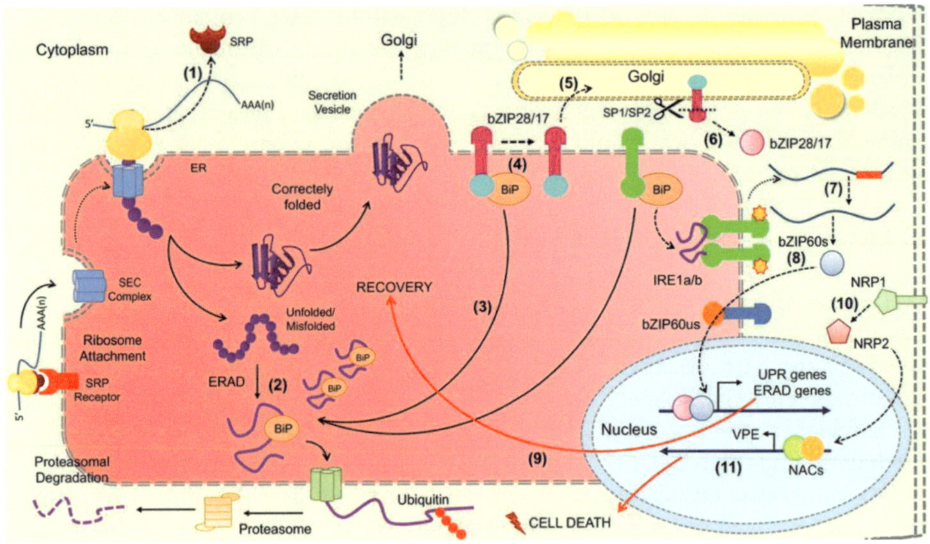

Figure 1.
The endoplasmic reticulum stress response in Arabidopsis. *The secretory proteins are synthesized in ER-bound polysomes (1) attached to the ER membrane through the interaction of signal recognition particle (SRP) and membrane receptor. As soon as they enter the lumen of the organelle, they are bound to a series of molecular chaperones, including BiP, to assist correct folding (2). Upon ER stress, the accumulation of unfolded protein (UP) activates a protective signaling cascade, designated as unfolded protein response, which allows communication of ER with the nucleus via a bipartite signaling module: the bZIP28/bZIP17 and IRE1a/IRE1b-bZiP60 signaling modules. Under normal conditions, BiP is bound to bZIP28/17, keeping the transducer in an inactive configuration (4). Upon ER stress, UP causes the dissociation of BiP from bZIP28/bZIP17, which is, then, translocated to the Golgi (5), where it is proteolytically cleaved to release the bZIP28/bZIP17 domain from the membrane that, in turn, is translocated to the nucleus (6). UP accumulation also causes the oligomerization of IRE1a/IRE1b, subsequent activation of its kinase domain by phosphorylation, and the endonuclease activity (6). The activated IRE1a/IRE1b endonuclease domain promotes unconventional splicing of bZIP60 mRNA to remove a transmembrane motif-encoding fragment, generating bZIP60 spliced mRNA that is translated into a soluble bZIP60 protein (bZIP60s) (7), which otherwise would be translated into the membrane-associated bZIP60us as it occurs under normal conditions. bZIP60s is, then, translocated to the nucleus (8), where it cooperates with bZIP28/bZIP17 to upregulate UPR genes and ERAD-related genes, increasing the ER protein processing capacity under ER stress to promote recovery (9). However, if the stress persists, and ER homeostasis cannot be restored, cell death signaling pathways are activated. among them, the DCD/NRP-mediated cell death signaling is initiated with activation of AtNRP1 (10) that leads to the induction of AtNRP2 and activation of a signaling cascade that culminates with the induction of ANAC36 that binds to the* VPE *promoter (11) and induces the expression of* VPE, *the executioner of the cell death program via collapse of the vacuole. These ER stress signaling pathways are conserved in other plant species.*

This unconventional splicing seems to prevent the degradation of XBP1s (spliced form) product by the proteasome and increase its transactivation activity, causing activation of UPR-related genes [16, 17]. Thus, the XBP1s is a soluble and functional transcription factor, which is reallocated to the cell nucleus to activate genes involved in cytoprotective pathways, such as some members of ER quality control or programmed cell death-related genes, including the apoptotic signaling kinase 1 (ASK1) and Jun-N-terminal kinase (JNK) [16–19].

The ER signal transducer ATF6 is anchored to the ER membrane and harbors an N-terminal sensor domain facing the ER lumen and a C-terminal bZIP domain facing the cytosolic side. Under normal conditions, ATF6 is inactivated by BiP binding to the ER stress sensor domain. ER stress conditions promote the BiP disassociation and reallocation of ATF6 to the Golgi apparatus, where it is specifically processed by SP1 and SP2 proteolytic enzymes. The limited proteolysis of ATF6 transmembrane domain allows that the bZIP domain of ATF6 be directed to the nucleus, where it acts in concert with XBP1 to induce genes involved in ER protein processing, ER quality control, and ER-associated protein degradation (ERAD) pathway.

Finally, the PERK activation upon BiP release by stress conditions promotes global translation suppression through the phosphorylation of the translation initiation factor IF2α [20]. PERK also activates the transcription factor CHOP, involved in the regulation of apoptosis-related genes [10, 21].

In plants, the UPR pathway has, at least, two arms (**Figure 1**). The first one activates IRE1 (IRE1a–AT2G17520 and IRE1b–AT5G24360, in *Arabidopsis thaliana*), and the other is transduced through bZIP membrane-associated transcription factors (bZIP17–AT2G40950 and bZIP28–AT3G10800, in *Arabidopsis thaliana*) [22, 23]. In the first arm of plant UPR, like in mammals, the accumulation of misfolded proteins leads to the activation of IRE1, which promotes unconventional cytosolic splicing of bZIP60 mRNA [24]. The unspliced bZIP60 mRNA, called bZIP60us, is translated into an ER membrane-associated transcription factor and does not exhibit transcriptional activity. Upon IRE1 activation by UPR, the spliced bZIP60 mRNA, called bZIP60s, does not display the transmembrane domain coding region, and its translation generates an active transcription factor, which is reallocated to the nucleus to activate UPR and cytoprotective genes, such as *BiP3*, *CNX* (calnexin), *CRT* (calreticulin), etc. [24–26]. This mechanism is conserved among plants, as the rice (*Oryza sativa*) bZIP60 orthologs, OsbZIP74 or OsbZIP50, display similar IRE-mediated mRNA splicing to render the activation of ER stress-inducible promoters [27, 28]. Likewise, in maize (*Zea mays*), ZmbZIP60 mRNA splicing leads to the activation of ER stress-inducible promoters [29], and, in soybean (*Glycine max*), the ZIP60 ortholog GmbZIP68 harbors a canonical site for IRE1 endonuclease activity and is efficiently spliced under ER stress conditions to activate UPR genes [30].

The second arm of plant UPR pathway is mediated by posttranslational modification of bZIP17 and bZIP28 transcription factors, the functional analogs of ATF6. Both bZIP17 and bZIP28 display a canonical SP1 site in their C-terminal domain, facing the ER lumen [31]. Upon stress conditions, BIP is released from the bZIP28 and bZIP17 ER sensor domain, and the transcription factors are reallocated from the ER to the Golgi apparatus, where they are processed by SP1 and SP2 proteases. These proteases remove the transmembrane domain of bZIP17 and bZIP28, exposing their cytosolic regions, which will activate UPR-related genes in the nucleus [31–34]. Like the IRE1/bZIP60 signaling module of plant UPR, the bZIP28/bZIP17 arm triggers the evolutionarily conservative UPR but also accommodates cross-talk with several other adaptive signaling responses [24, 30, 31]. In summary, upon ER stress, bZIP60s and bZIP28 use a different mechanism to be translocated to the nucleus where they act in concert to induce the expression of UPR genes and ERAD-related genes to increase the ER protein processing capacity for recovery from stress.

2.3 Convergence of ER stress and osmotic stress responses into a cell death signaling pathway

At a physiological level, the UPR encompasses three protective mechanisms: (i) global translation suppression by PERK-mediated IF2α phosphorylation; (ii) upregulation of ER-resident molecular chaperones, and (iii) proteasome-mediated protein degradation by ERAD pathway. However, if the stress conditions are sustained and the UPR pathway fails to restore ER homeostasis, apoptotic pathways are triggered as an ultimate attempt to survive. In plants, there is a specific branch of ER stress that integrates the osmotic stress and leads to programmed cell death (PCD), the development and cell death domain-containing N-rich protein (DCD/NRP)-mediated cell death signaling (**Figure 1**) [12]. This cell death pathway was first identified via genome-wide and expression profiling approaches, which revealed a modest overlapping between ER and osmotic stress-induced transcriptomes of soybean seedlings treated with PEG (an osmotic stress inducer) and tunicamycin and AZC (ER stress inducers). Several genes displayed similar kinetics and a synergistic induction under combined ER and osmotic stresses, indicating that the ER stress response integrates the osmotic signal to potentiate transcription of shared target genes. Among them, two plant-specific DCD/N-rich proteins, NRP-A and NRP-B, an ubiquitin-associated protein homolog (UBA), and a NAC domain-containing protein, GmNAC81, displayed the most robust synergistic upregulation by the combination of both stresses [35]. Transient expression of NRPs or GmNAC81 in soybean protoplasts and *Nicotiana benthamiana* leaves demonstrated that they are critical mediators of ER stress- and osmotic stress-induced cell death in plants [36–38].

The NRP-A and NRP-B display a highly conserved DCD domain at their C-terminal protein region and a high number of asparagine residues at their more divergent N-terminus (**Figure 2**) [39]. Consistent with the presence of a DCD domain, overexpression of *NRPs* in soybean protoplasts induces caspase-3-like activity and promotes extensive DNA fragmentation. Furthermore, transient

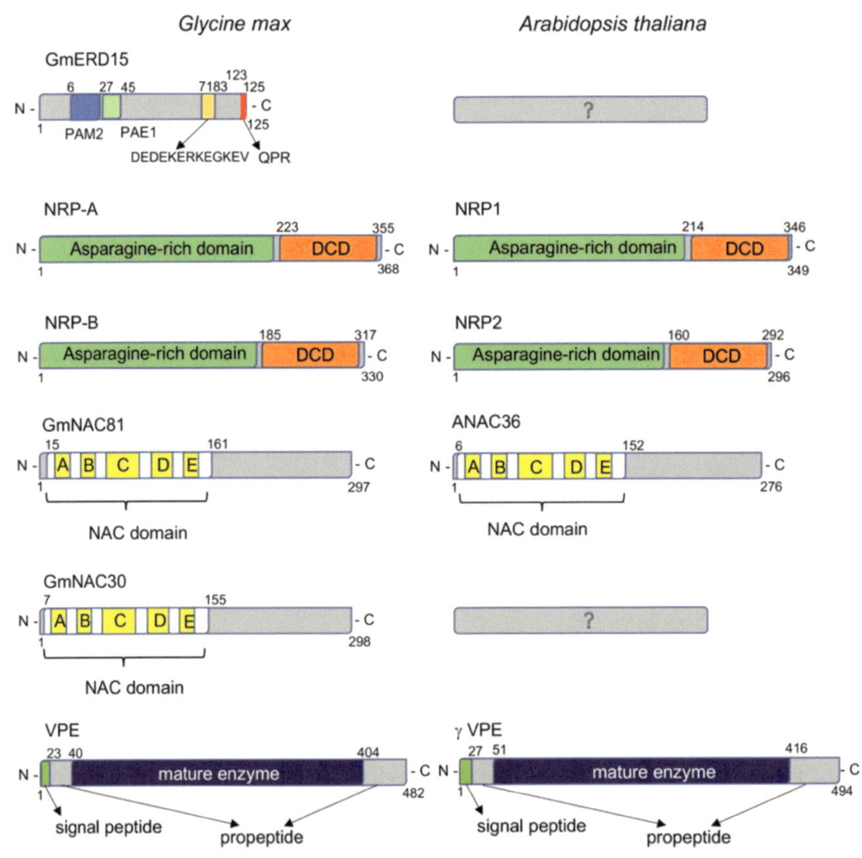

Figure 2.
Schematic representation of the cell death pathway components. The predicted domains of each protein are highlighted. The indicated domains are delimited by the amino acid positions in the primary structure shown by the numbers. For ERD-15, PAM2 is a PABP-interacting motif, PAE2 is PAM2-associated element 1 motif, DEDEKERKEGKEV is a conserved sequence representing a putative motif of ssDNA-binding transcriptional regulators, and QPR is a highly conserved C-terminal QPR motif. As for GmNAC81, GmNAC30, and ANAC36, the N-terminal NAC domain is subdivided into five conserved motifs (A to E) as indicated. In the AtNRP1, AtNRP2, NRP-A, and NRP-B schemes, DCD is development and cell death domain.

expression of NRPs *in planta* causes leaf yellowing, chlorophyll loss, malondialdehyde production, ethylene evolution, and induction of the senescence marker genes, which are hallmarks of leaf senescence and cell death [36, 38, 40]. The cell death response mediated by NRPs resembles a programmed cell death event. Because NRPs were the first components of the ER stress and osmotic stress-integrating cell death response to be characterized, this signaling pathway is commonly referred to as the DCD/NRP-mediated cell death response.

Similar to NRPs, GmNAC81 (*Glycine max* NAC81, formerly designated as GmNAC6) is another target of the ER stress- and osmotic stress-integrating pathway that induces a senescence-like response *in planta* and cell death in soybean protoplasts [37, 41]. GmNAC81 belongs to the plant-specific transcriptional factor superfamily of domain-containing proteins, represented by 111 members in *Arabidopsis*, 151 in rice, 152 in maize, and 180 in soybean [42, 43]. Members of this family function in development and stress response. The NAC transcriptional factors display a highly conserved N-terminal domain, called NAC domain, responsible for recognition of cis-regulatory elements on target promoters and DNA binding (**Figure 2**). The C-terminal domain is more divergent in sequence but is undoubtedly responsible for transcriptional activity [44, 45]. In addition, a subset of NAC proteins, which also exhibits protein binding activity, harbors an additional transmembrane domain present in the membrane-tethered NAC proteins [43, 46, 47].

NRPs and *GmNAC81* are induced by several different abiotic and biotic stresses in a coordinated manner, but induction of *NRPs* precedes the upregulation of *GmNAC81*. This early induction kinetics of *NRPs* is consistent with its capacity to activate the promoter and induce the expression of *GmNAC81*. These data placed GmNAC81 downstream of NRPs in the ER and osmotic stress-induced cell death pathway [37]. More recently, using reverse genetics in *Arabidopsis*, NRPs were confirmed to be upstream of ANAC36, the *Arabidopsis* ortholog of GmNAC81, in the DCD/NRP-mediated cell death signaling [40].

3. Early dehydration responsive gene 15, ERD15-like, controls NRP expression

The early dehydration responsive (*ERD*) genes were first identified due to their rapid induction in response to drought stress. The ERD genes (*ERD1* to *ERD16*) encode a set of proteins that differ in biological functions and cell localization [48]. Among them, ERD15 is a small acidic and hydrophilic protein that belongs to the PAM2 domain-containing protein family (**Figure 2**). The PAM2 domain is a well-characterized protein–protein interaction domain, which allows ERD15 to interact with polyA-binding proteins (PABP) regulating mRNA stability and protein translation [49]. In addition to PAM2, ERD15 contains two other domains with unknown function, designated as PAM2-associated element 1 (PAE1) and QPR.

ERD15 is a multiple stress-responsive gene that is involved in adaptation to abiotic and biotic stress. Light treatment, cold stress, and high salinity trigger *ERD15* expression [50, 51]. ERD15 functions as a negative regulator of the abscisic acid (ABA)-mediated response and a positive regulator of the salicylic acid (SA)-dependent defense pathway. *ERD15*-overexpressing transgenic lines are less sensitive to ABA and display enhanced salicylic acid-dependent defense pathway, which was associated with increased resistance to the bacterial *Erwinia carotovora* of the transgenic lines [52].

Consistent with the multiple stress-responsive expression profiles, the soybean *ERD15* ortholog (*GmERD15*) is also induced by ER and osmotic stress. *GmERD15* was identified using one hybrid screening that targeted the NRP-B promoter in yeast. As an upstream member of the NRP-mediated cell death response, GmERD15 binds the *NRP-B* promoter region in vivo and in vitro and induces the *NRP-B* expression [53]. Despite its role as a transcription factor, GmERD15 does not harbor a typical DNA-binding motif, but instead, it contains a conserved sequence of 13 amino acids at positions 71–83 (DEDEKERKEgKEv), which is a part of a tripartite motif domain derived from ssDNA-binding transcriptional regulators [54]. Accordingly, the GmERD15 binding site was mapped to a 12-bp palindromic sequence $^{-511}$AGCAnnnnnTGCT^{-500} on the *NRP-B* promotor in both single-stranded and double-stranded configurations [53].

4. The stress-induced NRP/NAC081/VPE module transduces a cell death signal

As components of the DCD/NPR-mediated cell death signaling, NRPs and GmNAC81 are critical mediators of cell death derived from ER stress and osmotic stress signals. More recent progress toward deciphering this branch of stress-induced cell death signaling includes the identification of two additional down-stream components, the NAC transcriptional factor (GmNAC30) and the vacuolar processing enzyme (VPE) [55].

GmNAC30 was identified as a nuclear partner of GmNAC81 via two-hybrid screening using GmNAC81 as a bait. *GmNAC30* and *GmNAC81* exhibit similar expression profiles and cell death activity. They are upregulated by ER stress, osmotic stress, and by the cell death-inducer cycloheximide. Consistently, GmNAC30 promotes cell death when transiently expressed in soybean protoplasts and, as a downstream component of the cell death signaling, is induced by expression of NRP-A and NRP-B.

GmNAC30 interacts with GmNAC81 in vitro and in vivo, the complex formed binds to common cis-regulatory sequences in target promoters and synergistically regulates hydrolytic enzyme promoters, including the caspase-1-like vacuolar processing enzyme (*VPE*) gene, which is involved in PCD in plants [55]. Consistent with their transcriptional function as a heterodimer, *GmNAC81* and *GmNAC30* display overlapping and coordinate expression profiles in response to multiple environmental and developmental stimuli. Therefore, the stress-induced *GmNAC30* cooperates with *GmNAC81* to activate PCD through the upregulation of the cell death executioner VPE.

VPE is a vacuole-localized cysteine protease that exhibits caspase-1-like activity and hydrolyzes a peptide bond at the C-terminal side of aspartate and asparagine residues [56]. It is synthesized as an inactive preprotein precursor, which is self-catalytically converted into the active mature form, under a processing step that resembles the activation of caspase 1 (**Figure 2**). It has been associated with *Tobacco mosaic virus*-induced hypersensitive cell death and developmental PCD [57, 58]. As an executioner of a cell death program, VPE is self-activated by hydrolytic cleavage and, in turn, mediates the initial activation of vacuolar enzymes, which degrade the vacuolar membrane and initiate the proteolytic cascade leading to PCD. Therefore, VPE activation may result in vacuolar collapse-mediated cell death, a type of plant-specific programmed cell death.

The discovery of VPE as a downstream target of the coordinate action of GmNAC81 and GmNAC30 underlies a mechanism for the execution of the ER and osmotic stress-induced cell death program (**Figure 1**). This model holds that prolonged ER and osmotic stresses induce the expression of the transcriptional activator GmERD15 to target the NRP promoter. The upregulation of NRPs initiates a transduction signaling that leads to the induction of GmNAC81 and GmNAC30, which cooperate to activate the VPE promoter and expression. Activation of VPE promotes the disintegration of vacuoles, initiating the proteolytic cascade in plant PCD. As vacuole-triggered PCD is unique to plants, the regulatory circuit linking the stress signal to activation of VPE is fundamentally composed of plant-specific signaling components.

The DCD/NRP-mediated programmed cell death pathway is conserved and operates with similar regulatory mechanisms in plants [40]. Soybean prototypes of each component of the cell death pathway were used to search for orthologs in the *Arabidopsis* genome (**Figure 3**) [30]. Arabidopsis AtNRP1 is most closely related to GmNRP-A and GmNRP-B, whereas a third homolog GmNRP-C was

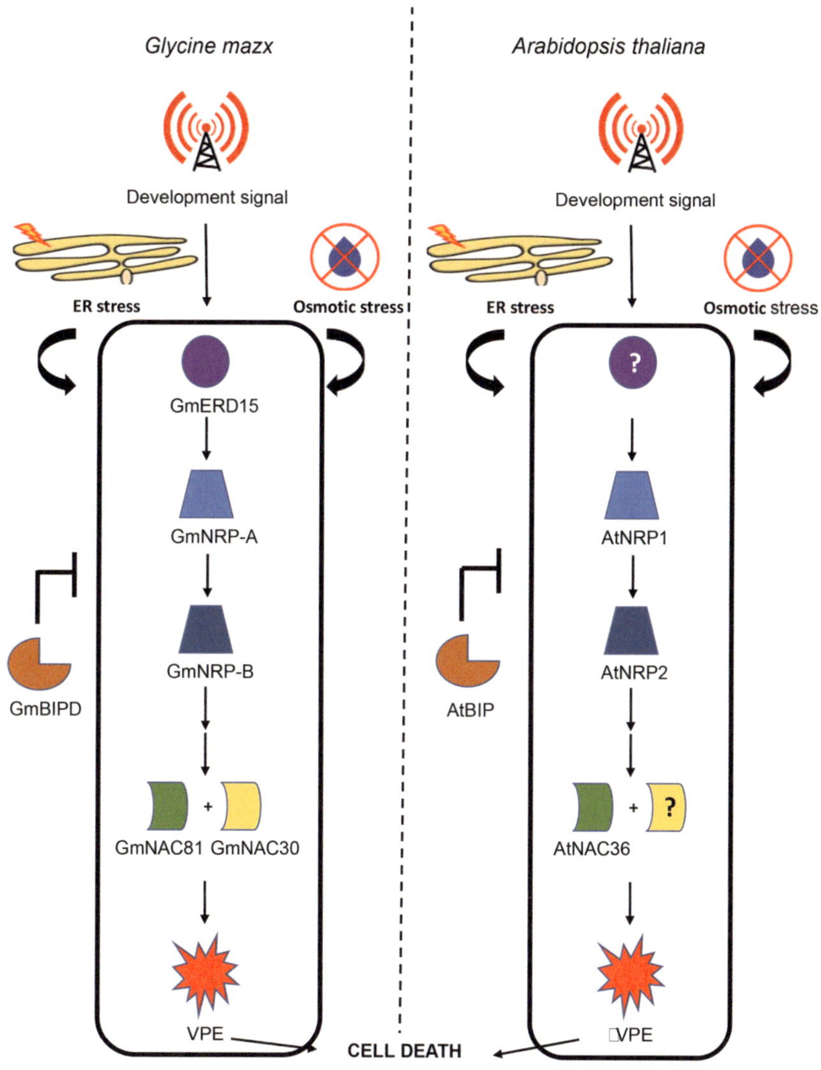

Figure 3.
Integration of developmental signal and stress signals into the DCD/NRP-mediated cell death response. Leaf senescence, ER stress, and osmotic stress induce the expression of ERD15-regulated NRP-A that in turn upregulates NRP-B to initiate a signaling cascade that culminates with the induction of GmNAC30 *and* GmNAC81 *expression. The NAC transcription factors form a heterodimer to fully induce the activation of* VPE *promoter, which leads to* VPE *upregulation and subsequent execution of a cell death program. The ER-resident molecular chaperone BiP acts as a negative regulator of cell death by modulating the expression and activity of the cell death pathway components. The DCD/NRP-mediated cell death signaling is conserved in other plant species, and the* Arabidopsis *orthologs are shown on the right.*

related to AtNRP-2. GmNAC81 and its paralog share sequence conservation with the *Arabidopsis* ortholog ANAC36 (At2G17040), whereas the predicted *Arabidopsis* ortholog of soybean VPE was identified as At4G32940/γVPE. Transient expression of the selected *Arabidopsis* orthologs of pathway components (*AtNRP-1*, *AtNRP-2*, *ANAC36*, and *γVPE)* induces cell death in *Nicotiana benthamiana* leaves with the appearance of hallmarks of PCD and leaf senescence, including DNA fragmentation, leaf yellowing, chlorophyll loss, and lipid peroxidation [38]. In addition, knockout lines for each one of pathway genes in *Arabidopsis* display enhanced tolerance to ER stress-mediated cell death induction. Very importantly, the stress induction of *AtNRP2*, *ANAC36*, and *γVPE* was dependent on the AtNRP1 function, confirming the upstream position of AtNRP1 in the cell death pathway. Therefore, in *Arabidopsis*, the execution of the cell death program has been proposed to occur through AtNRP1-mediated induction of the AtNRP2-ANAC36-γVPE signaling module. Nevertheless, functional information about the GmERD15 and GmNAC30 orthologs in *Arabidopsis* is lacking, and these pathway components have not been identified yet in *Arabidopsis*. Both in soybean and *Arabidopsis*, the DCD/NRP-mediated cell death pathway is modulated by the ER-resident molecular chaperone BiP, which negatively regulates the gene expression and activity of these cell death-inducing genes [13, 40].

5. A negative regulator of the NRP/NAC081/VPE signaling module confers tolerance to drought

Plants can negatively modulate the NRP/DCD-mediated cell death response to suit the cellular balance during the stress conditions. Moreover, this modulation improves the cellular stableness and consequently increases the plant tolerance to stress conditions in an essential process that is required for plant acclimatization and development. The molecular chaperone BiP plays a crucial role as a negative regulator of NRP/DCD-mediated cell death response. BiP belongs to the HSP70 family, which is essential to protect the cells against environmental stresses and to restore the cell homeostasis [59].

The molecular chaperone BiP has a catalytic site at the amino-terminal region and a substrate-binding site at the carboxy-terminal region [60]. BiP is involved in the regulation of several processes in the endoplasmic reticulum, a critical organelle that is related to responses to abiotic and biotic stress in plants. In the ER, BiP acts as a sensor that responds to quantitative and qualitative changes in the ER by regulating the activity of ER stress transducers [61]. Furthermore, BiP coordinately regulates the cell death signaling, which connects the signals from osmotic and ER stress in a DCD/NRP-dependent manner [35, 36, 38].

BiP attenuates the NRP/DCD-mediated cell death signal propagation by the modulation of expression and activity of the pathway signaling components (**Figure 3**). BiP overexpression in soybean attenuates ER stress- and osmotic stress-mediated cell death, a phenotype that is linked to a delay in the induction of *GmNRP-A*, *GmNRP-B*, and *GmNAC81* under ER stress and osmotic stress [38]. Furthermore, enhanced accumulation of BiP in tobacco (*Nicotiana tabacum*) prevents the GmNRP- and GmNAC81-mediated induction of cell death-associated physiological and molecular markers, whereas silencing of endogenous BiP enhances the cell death response.

In addition to alleviating ER and osmotic stress-mediated cell death, the *BiP* overexpression in plants has also been shown to increase their tolerance to water deficits [62–64]. Enhanced accumulation of BiP in soybean, tobacco, and *Arabidopsis* promotes a delay in drought-induced senescence and wilting of leaves leading to a higher survival rate of overexpressing lines under water-deficit regimes [12, 38, 40, 63–64]. The BiP-mediated tolerance mechanism is not associated with conventional mechanisms of drought tolerance and avoidance, as the BiP-overexpressing lines do not display lower photosynthesis and transpiration rates than untransformed lines under drought, and the stomata closure and root growth are not stimulated under water deprivation. Furthermore, the *BiP*-overexpressing lines exhibit a lower induction of drought-related genes than WT under water-deficit conditions, and the abscisic acid content in *BiP*-overexpressing plants is similar to untransformed lines, indicating that the BiP-mediated drought tolerance mechanism is independent on ABA [59, 64, 65]. Under drought conditions, the only variations observed in *BiP*-overexpressing lines are a delay in drought-induced leaf senescence and an attenuation in the drought induction of PCD-associated marker genes, which is associated with the protective function of BiP as a negative modulator of the DCD/NRP-mediated cell death response. A metabolomic approach was used to detect the metabolite profile of *BiP*-overexpressing lines under drought conditions [65]. Due to a higher osmolyte accumulation, mainly amino acids, the *BiP*-overexpressing plants can maintain the leaf turgidity upon drought stress, which is a phenotypic hallmark of the BiP-mediated tolerance to drought. The *BiP*-overexpressing lines also display a higher accumulation of salicylic acid and upregulation of SA-responsive genes, which is associated with accelerated hypersensitive response triggered by *Pseudomonas syringae pv tomato* in soybean and tobacco [59, 65]. The SA signaling also activates the antioxidative metabolism, which may be linked to the BiP protective function to drought. Very importantly, the BiP modulation of the DCD/NRP-mediated cell death response does not impair the plant growth and development.

6. The stress-induced DCD/NRP-mediated cell death signaling positively regulates leaf senescence

Leaf senescence is a natural process in plant development, which begins with a physiological transition between active photosynthetic leaves to degenerative and nutrient-recycling leaves. The classical age senescence-related symptom is the leaf yellowing caused by generalized chlorophyll loss. The age-induced senescence or naturally programmed leaf senescence, hereafter referred to as leaf senescence, occurs by plant aging and is precisely regulated by senescence-associated genes (SAGs) [66, 67].

Many SAGs are environmental- and stress-responsive genes, integrating a convergent regulatory cascade between natural plant development and stress-induced PCD [68]. At the molecular level, the onset of senescence is accompanied by a massive reprogramming of gene expression, probably controlled by senescence-associated transcription factors. Among these, several NAC transcription factors have been associated with senescence regulation based on high-resolution temporal expression profiles [69].

In soybean, a transcriptomic analysis of senescing leaves reveals that 44% of the *GmNAC* genes were differentially expressed at the onset of leaf senescence. The most representative subfamilies of soybean senescence-associated *NAC* genes were the abiotic stress-induced SNAC-A (ATAF) subfamily, in which 90% of the members were differentially expressed during senescence, followed by the biotic stress-induced TERN subfamily, displaying 80% of the members differentially expressed during leaf senescence [43]. *GmNAC30* and *GmNAC81*, which belong to the SNAC-A and TERN subfamilies, respectively, are among the upregulated genes by leaf senescence [43, 59]. These results raise the hypotheses that the (i)

DCD-NRP/NAC/VPE signaling module may integrate stress-induced with natural leaf senescence and (ii) other NAC genes may be involved in integrated circuits between age- and stress-induced cell death pathways.

Regarding the first hypothesis, several lines of evidence indicate that the regulatory circuit NRPs/GmNAC81:GmNAC30/VPE integrates osmotic stress- and ER stress-induced PCD response with natural leaf senescence. First, not only *GmNAC30* and *GmNAC81* but also the other cell death pathway components, *NRP-A*, *NRP-B*, and *VPE*, are induced by leaf senescence [43, 59, 70]. Second, the activity of VPE is also induced during the onset of leaf senescence [59]. Third, transient expression of the soybean components of ER stress- and osmotic stress-induced cell death response, NRP-A, NRP-B, GmNAC81, and GmNAC30, as well as the *Arabidopsis* orthologs AtNRP1, AtNRP2, ANAC36, and γVPE, in protoplasts and *in planta* induce a cell death response bearing the hallmarks of leaf senescence and PCD. These symptoms include the induction of caspase 1-like activity and DNA fragmentation, chlorophyll loss, protein degradation, enhanced lipid peroxidation, and the induction of senescence-associated marker genes [36–38, 40, 55]. Fourth, enhanced accumulation of BiP, which negatively regulates the NRPs/GmNAC81:GmNAC30/VPE signaling module, also promotes a delay in leaf senescence in transgenic plants [59]. Finally, GmNAC81 is a positive regulator of naturally programmed leaf senescence [70]. Although leaf senescence is genetically programmed in an age-dependent manner, it can be triggered by environmental cues and is also positively and negatively regulated by various plant hormones. *GmNAC81* and *GmNAC30* are induced by the phytohormones ABA, jasmonic acid (JA) and salicylic acid (SA), which are positive regulators of senescence, and GmNAC81-overexpressing lines display high levels of ABA, mimicking the enhanced endogenous levels of this hormone during leaf senescence [70, 71]. Consistent with a role in leaf senescence, the overexpression of *GmNAC81* in soybean plants accelerates leaf senescence, a phenotype associated with extensive leaf yellowing, increased chlorophyll loss, faster photosynthetic decay, and enhanced expression and activity of the GmNAC81 direct target VPE, than untransformed, wild-type plants. Conversely, suppressing *GmNAC81* expression delays leaf senescence and decreases the expression of GmNAC81 direct target genes, including *VPE* [70]. Therefore, GmNAC81 is involved in developmentally programmed leaf senescence. Furthermore, ER stress- and osmotic stress-induced PCD is integrated with natural leaf senescence through the NRPs/NACs/VPE regulatory circuit.

7. Conclusion

Since the discovery of the ER stress- and osmotic stress-induced DCD/NRP-mediated cell death response, considerable progress has been achieved toward deciphering the components and regulation of the pathway (**Figure 3**). We now know that the combination of multiple stresses synergistically activates a plant-specific PCD response that is initiated by induction of the stress-responsive transcription factor GmERD15, which, in turn, binds and activates the DCD/NRP promoter. Induction of the DCD/NRP genes *NRP-A* and *NRP-B* leads to the activation of a signal cascade that culminates with the upregulation of the transcription factors GmNAC81 and GmNAC30. The NAC transcription factors form a heterodimer to activate the expression of hydrolytic enzymes, including VPE, an executioner of vacuole-triggered programmed cell death. The stress-induced DCD/NRP-mediated cell death response is conserved in plants with similar regulatory mechanisms and represents a shared response to multiple stress signals. As a negative regulator of the stress-induced DCD/NRP-mediated cell death response, overexpression of the

ER-resident molecular chaperone BiP delays drought-induced senescence in tobacco and soybean plants and confers the increased adaptation of these transgenic lines under water deprivation conditions. This DCD/NNP-mediated stress-induced cell death program is also activated during age-dependent leaf senescence and contributes positively for the progression of the developmentally programmed senescence. Therefore, the plant-specific NRPs/NACs/VPE signaling module represents a regulatory circuit integrating stress-induced with natural leaf senescence.

Acknowledgements

We thank the Brazilian government agencies, Conselho Nacional de Desenvolvimento Científico e Tecnológico (CNPq), Fundação de Amparo à Pesquisa do Estado de Minas Gerais (FAPEMIG), and Coordenação de Aperfeiçoamento de Pessoal de Nível Superior (CAPES), for financial support.

Conflict of interest

The authors declare no conflict of interest.

Author details

Otto Teixeira Fraga[1,2], Bruno Paes de Melo[1,2], Luiz Fernando de Camargos[1,2], Debora Pellanda Fagundes[1], Celio Cabral Oliveira[1,2], Eduardo Bassi Simoni[1,2], Pedro Augusto Braga dos Reis[1,2] and Elizabeth Pacheco Batista Fontes[1,2]*

1 National Institute of Science and Technology in Plant-Pest Interactions, Bioagro, Universidade Federal de Viçosa, Viçosa, Brazil

2 Department of Biochemistry and Molecular Biology/BIOAGRO, Universidade Federal de Viçosa, Viçosa, Brazil

*Address all correspondence to: bbfontes@ufv.br

References

[1] Pandey S. Heterotrimeric G-protein signaling in plants: Conserved and novel mechanisms. Annual Review of Plant Biology. 2019;**70**:213-238. DOI: 10.1146/annurev-arplant-050718-100231

[2] Mahajan S, Tuteja N. Cold, salinity and drought stresses: An overview. Archives of Biochemistry and Biophysics. 2005;**444**:139-158. DOI: 10.1016/j.abb.2005.10.018

[3] Zandalinas SI, Mittler R, Balfagóna D, Arbonaa V, Gómez-Cadenas A. Plant adaptations to the combination of drought and high temperatures. Physiologia Plantarum. 2018;**162**:2-12. DOI: 10.1111/ppl.12540

[4] Stone SL. Role of the ubiquitin proteasome system in plant response to abiotic stress. International Review of Cell and Molecular Biology. 2019;**343**:65-110. DOI: 10.1016/bs.ircmb.2018.05.012

[5] Mansfield TA, Atkinson CJ. Stomatal behaviour in water stressed plants. In: Alscher RG, Cumming JR, editors. Stress Responses in Plants : Adaptation and Acclimation Mechanisms. New York: Wiley-Liss, Inc.; 1990. pp. 241-264

[6] Cheng WH, Endo A, Zhou L, Penney J, Chen HC, Arroyo A, et al. A unique short-chain dehydrogenase/reductase in Arabidopsis glucose signaling and abscisic acid biosynthesis and functions. The Plant Cell. 2002;**14**:2723-2743. DOI: 10.1105/tpc.006494

[7] Bota J, Medrano H, Flexas J. Is photosynthesis limited by decreased Rubisco activity and RuBP content under progressive water stress? New Phytologist. 2004;**162**:671-681. DOI: 10.1111/j.1469-8137.2004.01056.x

[8] Baxter A, Mittler R, Suzuki N. ROS as key players in plant stress signalling. Journal of Experimental Botany. 2014;**65**:1229-1240. DOI: 10.1093/jxb/ert375

[9] Zhu JK. Abiotic stress signaling and responses in plants. Cell. 2016;**167**:313-324. DOI: 10.1016/j.cell.2016.08.029

[10] Eichmann R, Schafer P. The endoplasmic reticulum in plant immunity and cell death. Frontiers in Plant Science. 2012;**3**:200. DOI: 10.3389/fpls.2012.00200

[11] Hetz C. The unfolded protein response: Controlling cell fate decisions under ER stress and beyond. Nature Reviews Molecular Cell Biology. 2012;**12**:89-112. DOI: 10.1038/nrm3270

[12] Reis P, Fontes EPB. N-rich protein (NRP)-mediated cell death signaling: A new branch of the ER stress response with implications for plant biotechnology. Plant Signaling & Behavior. 2012;7:628-632. DOI: 10.4161/psb.20111

[13] de Camargos LF, Fraga OT, Oliveira CC, da Silva JCF, Fontes EPB, Reis PAB. Development and cell death domain-containing asparagine-rich protein (DCD/NRP): An essential protein in plant development and stress responses. Theoretical and Experimental Plant Physiology. 2019;**31**:59-70. DOI: 10.1007/s40626-018-0128-z

[14] Schwarz DS, Blower MD. The endoplasmic reticulum: Structure, function and response to cellular signaling. Cellular and Molecular Life Sciences. 2016;**73**:79-94. DOI: 10.1007/s00018-015-2052-6

[15] Water P, Ron D. The unfolded protein response: From stress pathway to homeostatic regulation. Science. 2011;**334**:1081-1086. DOI: 10.1126/science.1209038

[16] Yoshida H, Matsui T, Yamamoto A, Okada T, Mori K. XBP1 mRNA is induced by ATF6 and spliced by IRE1 in response to ER stress to produce a highly active transcription factor. Cell. 2001;**107**:881-891. DOI: 10.1016/s0092-8674(01)00611-0

[17] Chen CY, Malchus NS, Hehn B, Stelzer W, Avci D, Langosch D, et al. Signal peptide peptidase functions in ERAD to cleave the unfolded protein response regulator XBP1u. EMBO Journal. 2014;**33**:2492-2506. DOI: 10.15252/embj.201488208

[18] Koizumi N, Martinez IM, Kimata Y, Kohno K, Sano H, Chrispeels MJ. Molecular characterization of two Arabidopsis Ire1 homologs, endoplasmic reticulum-located transmembrane protein kinases. Plant Physiology. 2001;**127**:949-962. DOI: https://doi.org/10.1104/pp.010636

[19] Humbert S, Zhong S, Deng Y, Howell SH, Rothstein SJ. Alteration of the bZIP60/IRE1 pathway affects plant response to ER stress in Arabidopsis thaliana. PLoS One. 2012;**7**:e39023. DOI: 10.1371/journal.pone.0039023

[20] Harding HP, Novoa I, Zhang Y, Zeng H, Wek R, Schapira M, et al. Regulated translation initiation controls stress-induced gene expression in mammalian cells. Molecular Cell. 2000;**6**:1099-1108. DOI: 10.1016/S1097-2765(00)00108-8

[21] Urade R. The endoplasmic reticulum stress signaling pathways in plants. BioFactors. 2009;**35**:326-331. DOI: 10.1002/biof.45

[22] Ruberti C, Kim SJ, Stefano G, Brandizzi F. Unfolded protein response in plants: One master, many questions. Current Opinion in Plant Biology. 2015;**27**:59-66. DOI: 10.1016/j.pbi.2015.05.016

[23] Bao Y, Howell SH. The unfolded protein response supports plant development and defense as well as responses to abiotic stress the unfolded protein response supports plant development and defense as well as responses to abiotic stress. Frontiers in Plant Science. 2017;**8**:344. DOI: 10.3389/fpls.2017.00344

[24] Deng Y, Humbert S, Liu JX, Srivastava R, Rothstein SJ, Howell SH. Heat induces the splicing by IRE1 of a mRNA encoding a transcription factor involved in the unfolded protein response in Arabidopsis. Proceedings of the National Academy of Sciences of the United States of America. 2011;**108**:7247-7252. DOI: 10.1073/pnas.1102117108

[25] Iwata Y, Fedoroff NV, Koizumi N. Arabidopsis bZIP60 is a proteolysis-activated transcription factor involved in the endoplasmic reticulum stress response. The Plant Cell. 2008;**20**:3107-3121. DOI: 10.1105/tpc.108.061002

[26] Nagashima Y, Ki M, Suzuki E, Shimada Y, Iwata Y, Koizumi N. Arabidopsis IRE1 catalyses unconventional splicing of bZIP60 mRNA to produce the active transcription factor. Scientific Reports. 2011;**1**:29. DOI: 10.1038/srep00029

[27] Lu SJ, Yang ZT, Sun L, Sun L, Song ZT, Liu JX. Conservation of IRE1-regulated bZIP74 mRNA unconventional splicing in rice (Oryza sativa L.) involved in ER stress responses. Molecular Plant. 2012;**5**:504-514. DOI: 10.1093/mp/ssr115

[28] Hayashi S, Takahashi H, Wakasa Y, Kawakatsu T. Identification of a cis-element that mediates multiple pathways of the endoplasmic reticulum stress response in rice. Plant Journal. 2013;**74**:248-257. DOI: 0.1111/tpj.12117

[29] Li Y, Humbert S, Howell SH. ZmbZIP60 mRNA is spliced in maize in response to ER stress. BMC Research

Notes. 2012;**5**:144. DOI: 10.1186/1756-0500-5-144

[30] Silva PA, Silva JCF, Caetano HDN, Machado JPB, Mendes GC, Reis PAB, et al. Comprehensive analysis of the endoplasmic reticulum stress response in the soybean genome: Conserved and plant-specific features. BMC Genomics. 2015;**16**:783. DOI: 10.1186/s12864-015-1952-z

[31] Che P, Bussell JD, Zhou W, Estavillo GM, Pogson BJ, Smith SM. Signaling from the endoplasmic reticulum activates brassinosteroid signaling and promotes acclimation. Science Signaling. 2010;**3**:ra69. DOI: 10.1126/scisignal.2001140

[32] Liu JX, Srivastava R, Che P, Howell SH. An endoplasmic reticulum stress response in Arabidopsis is mediated by proteolytic processing and nuclear relocation of a membrane-associated transcription factor, bZIP28. The Plant Cell. 2007;**19**:4111-4119. DOI: 10.1105/tpc.106.050021

[33] Gao H, Brandizzi F, Benning C, Larkin RM. A membrane-tethered transcription factor defines a branch of the heat stress response in Arabidopsis thaliana. Proceedings of the National Academy of Sciences of the United States of America. 2008;**105**:16398-16403. DOI: 10.1073/pnas.0808463105

[34] Sun L, Zhang SS, Lu SJ, Liu JX. Site-1 protease cleavage site is important for the ER stress-induced activation of membrane-associated transcription factor bZIP28 in Arabidopsis. Science China. Life Sciences. 2015;**58**:270-275. DOI: 10.1007/s11427-015-4807-6

[35] Irsigler AST, Costa MDL, Zhang P, Reis PAB, Dewey RE, Boston RS, et al. Expression profiling on soybean leaves reveals integration of ER- and osmotic-stress pathways. BMC Genomics. 2007;**8**:431. DOI: 10.1186/1471-2164-8-431

[36] Costa MDL, Reis PAB, Valente MAS, Irsigler AST, Carvalho CM, Loureiro ME, et al. A new branch of endoplasmic reticulum stress signaling and the osmotic signal converge on plant-specific asparagine-rich proteins to promote cell death. The Journal of Biological Chemistry. 2008;**283**:20209-20219. DOI: 10.1074/jbc.M802654200

[37] Faria JAQA, Reis PAB, Reis MTB, Rosado GL, Pinheiro GL, Mendes GC, et al. The NAC domain–containing protein, GmNAC6, is a downstream component of the ER stress–and osmotic stress–induced NRP–mediated cell–death signaling pathway. BMC Plant Biology. 2011;**11**:129. DOI: 10.1186/1471-2229-11-129

[38] Reis PAB, Rosado GL, Silva LAC, Oliveira LC, Oliveira LB, Costa MDL, et al. The binding protein BiP attenuates stress-induced cell death in soybean via modulation of the N-rich protein-mediated signaling pathway. Plant Physiology. 2011;**157**:1853-1865. DOI: 10.1104/pp.111.179697

[39] Tenhaken R, Doerks T, Bork P. DCD—A novel plant specific domain in proteins involved in development and programmed cell death. BMC Bioinformatics. 2005;**6**:169. DOI: 10.1186/1471-2105-6-169.

[40] Reis PAB, Carpinetti PA, Freitas PPJ, Santos EGD, Camargos LF, Oliveira IHT, et al. Functional and regulatory conservation of the soybean ER stress-induced DCD/NRP-mediated cell death signaling in plants. BMC Plant Biology. 2016;**16**:156. DOI: 10.1186/s12870-016-0843-z

[41] Pinheiro GL, Marques CS, Costa MDBL, Reis PAB, Alves MS, Carvalho CM, et al. Complete inventory of soybean NAC transcription factors: Sequence conservation and expression analysis uncover their distinct roles in stress response. Gene. 2009;**444**:10-23. DOI: 10.1016/j.gene.2009.05.012

[42] Shao H, Wang H, Tang X. NAC transcription factors in plant multiple abiotic stress responses: Progress and prospects. Frontiers in Plant Science. 2015;**6**:902. DOI: 10.3389/fpls.2015.00902

[43] Melo BP, Fraga OT, Silva JCF, Ferreira DO, Brustolini OJB, Carpinetti PA, et al. Revisiting the soybean GmNAC superfamily. Frontiers in Plant Science. 2018;**9**:1864. DOI: 10.3389/fpls.2018.01864

[44] Olsen AN, Ernst HA, Leggio LL, Skriver K. NAC transcription factors: Structurally distinct, functionally diverse. Trends in Plant Science. 2005;**10**:79-87. DOI: 10.1016/j.tplants.2004.12.010

[45] Nakashima K, Takasaki H, Mizoi J, Shinozaki K, Yamaguchi-Shinozaki K. NAC transcription factors in plant abiotic stress responses. Biochimica et Biophysica Acta. 2012;**1819**:97-103. DOI: 10.1016/j.bbagrm.2011.10.005

[46] Tran LSP, Quach TN, Guttikonda SK, Aldrich DL, Kumar R, Neelakandan A, et al. Molecular characterization of stress-inducible GmNAC genes in soybean. Molecular Genetics and Genomics. 2009;**281**:647-664. DOI: 0.1007/s00438-009-0436-8

[47] Seo PJ, Kim MJ, Park JY, Kim SY, Jeon J, Lee YH, et al. Cold activation of a plasma membrane-tethered NAC transcription factor induces a pathogen resistance response in Arabidopsis. Plant Journal. 2010;**61**:661-671. DOI: 10.1111/j.1365-313X.2009.04091.x

[48] Kiyosue T, Yamaguchi-Shinozaki K, Shinozaki K. Cloning of cDNAs for genes that are early-responsive to dehydration stress (ERDs) in Arabidopsis thaliana L.: Identification of three ERDs as HSP cognate genes. Plant Molecular Biology. 1994;**25**:791-798. DOI: 10.1007/bf00028874

[49] Wang X, Grumet R. Identification and characterization of proteins that interact with the carboxy terminus of poly(A)-binding protein and inhibit translation in vitro. Plant Molecular Biology. 2004;**54**:85-98. DOI: 10.1023/B:PLAN.0000028771.70969.6b

[50] Dunaeva M, Adamska I. Identification of genes expressed in response to light stress in leaves of Arabidopsis thaliana using RNA differential display. European Journal of Biochemistry. 2001;**268**:5521-5529. DOI: 10.1046/j.1432-1033.2001.02471.x

[51] Park MY, Chung MS, Koh HS, Lee DJ, Ahn SJ, Kim CS. Isolation and functional characterization of the Arabidopsis salt-tolerance 32 (AtSAT32) gene associated with salt tolerance and ABA signaling. Physiologia Plantarum. 2009;**135**:426-435. DOI: 10.1111/j.1399-3054.2008.01202.x

[52] Kariola T, Brader G, Helenius E, Li J, Heino P, Palva ET. Early responsive to dehydration 15, a negative regulator of abscisic acid responses in Arabidopsis. Plant Physiology. 2006;**142**:1559-1573. DOI: 10.1104/pp.106.086223

[53] Alves MS, Reis PAB, Dadalto SP, Faria JAQA, Fontes EPB, Fietto LG. A novel transcription factor, ERD15 (early responsive to dehydration 15), connects endoplasmic reticulum stress with an osmotic stress-induced cell death signal. The Journal of Biological Chemistry. 2011;**286**:20020-20030. DOI: 10.1074/jbc.M111.233494

[54] Desveaux D, Allard J, Brisson N, Sygusch J. A new family of plant transcription factors displays a novel ssDNA-binding surface. Nature Structural and Molecular Biology. 2002;**9**:512-517. DOI: 10.1038/nsb814

[55] Mendes GC, Reis PAB, Calil IP, Carvalho HH, Aragao FJL, Fontes EPB. GmNAC30 and GmNAC81 integrate the endoplasmic reticulum stress- and

osmotic stress-induced cell death responses through a vacuolar processing enzyme. Proceedings of the National Academy of Sciences of the United States of America. 2013;**110**:19627-19632. DOI: 10.1073/pnas.1311729110

[56] Hara-Nishimura I, Hatsugai N, Nakaune S, Kuroyanagi M, Nishimura M. Vacuolar processing enzyme: An executor of plant cell death. Current Opinion in Plant Biology. 2005;**8**:404-408. DOI: 10.1016/j.pbi.2005.05.016

[57] Hatsugai N, Kuroyanagi M, Yamada K, Meshi T, Tsuda S, Kondo M, et al. A plant vacuolar protease, VPE, mediates virus-induced hypersensitive cell death. Science. 2004;**305**:855-858. DOI: 10.1126/science.1099859

[58] Hatsugai N, Yamada K, Goto-Yamada S, Hara-Nishimura I. Vacuolar processing enzyme in plant programmed cell death. Frontiers in Plant Science. 2015;**6**:234. DOI: 10.3389/fpls.2015.00234

[59] Carvalho HH, Silva PA, Mendes GC, Brustolini OJB, Pimenta MR, Gouveia BC, et al. The endoplasmic reticulum binding protein BiP displays dual function in modulating cell death events. Plant Physiology. 2014;**164**:654-670. DOI: 10.1104/pp.113.231928

[60] DB MK. Structure and mechanism of 70-kDa heat-shock-related proteins. Advances in Protein Chemistry. 1993;**44**:67-98. DOI: 10.1016/S0065-3233(08)60564-1

[61] Srivastava R, Deng Y, Shah S, Rao AG, Howell SH. Binding protein is a master regulator of the endoplasmic reticulum stress sensor/transducer bZIP28 in Arabidopsis. The Plant Cell. 2013;**25**:1416-1429. DOI: 10.1105/tpc.113.110684

[62] Cascardo JCM, Almeida RS, Buzeli RAA, Carolino SMB, Otoni WC, Fontes EPB. The phosphorylation state and expression of soybean BiP isoforms are differentially regulated following abiotic stresses. The Journal of Biological Chemistry. 2000;**275**:14494-14500. DOI: 10.1074/jbc.275.19.14494

[63] Alvim FC, Carolino SMB, Cascardo JCM, Nunes CC, Martinez CA, Otoni WC, et al. Enhanced accumulation of BiP in transgenic plants confers tolerance to water stress. Plant Physiology. 2001;**126**:1042-1054. DOI: 10.1104/pp.126.3.1042

[64] Valente MAS, Faria JAQA, Soares-Ramos JRL, Reis PAB, Pinheiro GL, Piovesan ND, et al. The ER luminal binding protein (BiP) mediates an increase in drought tolerance in soybean and delays drought-induced leaf senescence in soybean and tobacco. Journal of Experimental Botany. 2009;**60**:533-546. DOI: 10.1093/jxb/ern296

[65] Coutinho FS, dos Santos DS, Lima LL, Vital CE, Santos LA, Pimenta MR, et al. Mechanism of the drought tolerance of a transgenic soybean overexpressing the molecular chaperone BiP. Physiology and Molecular Biology of Plants. 2019;**25**:457-472. DOI: 10.1007/s12298-019-00643-x

[66] Yoshida S. Molecular regulation of leaf senescence. Current Opinion in Chemical Biology. 2003;**6**:79-84. DOI: 10.1016/S1369526602000092

[67] Lim PO, Kim HJ, Nam HG. Leaf senescence. Annual Review of Plant Biology. 2007;**58**:115-136. DOI: 10.1146/annurev.arplant.57.032905.105316

[68] Balazadeh S, Siddqui H, Allu AD, Matallana-Ramirez LP, Caldana C, Mehrnia M, et al. A gene regulatory network controlled by the NAC transcription factor ANAC092/AtNAC2/ORE1 during salt promoted senescence.

Plant Journal. 2010;**62**:250-264. DOI: 10.1111/j.1365-313X.2010.04151.x

[69] Breeze E, Harrison E, McHattie S, Hughes L, Hickman R, Hill C, et al. High-resolution temporal profiling of transcripts during Arabidopsis leaf senescence reveals a distinct chronology of processes and regulation. The Plant Cell. 2011;**23**:873-894. DOI: 10.1105/tpc.111.083345

[70] Pimenta MR, Silva PA, Mendes GC, Alves JR, Caetano HDN, Machado JPB, et al. The stress-induced soybean NAC transcription factor GmNAC81 plays a positive role in developmentally programmed leaf senescence. Plant and Cell Physiology. 2016;**57**:1098-1114. DOI: 10.1093/pcp/pcw059

[71] Zhang H, Zhou C. Signal transduction in leaf senescence. Plant Molecular Biology. 2013;**82**:539-545. DOI: 10.1007/s11103-012-9980-4

Chapter 6

Leaf Senescence in Wheat: A Drought Tolerance Measure

Hafsi Miloud and Guendouz Ali

Abstract

The present study was conducted on the experimental site of INRAA, unit research of Setif. A set of 10 genotypes of durum wheat (*Triticum durum* Desf.) planted during four cropping seasons (2009–2013). The objectives of this study are to evaluate the performance of some durum wheat genotypes and tested the efficiency of using senescence parameters in screening under semi-arid conditions. The analysis of variance demonstrates significant effects of genotypes and years on the grain yield and senescence parameters. Based on the means comparison, the values of total mean grain yield (2009–2013) varied from 37.84 q/ha for Oued Zenati to 44.7 q/ha for Altar84 with general mean of 42.71 q/ha. The mean rankings based on the mean grain yield demonstrate that the genotypes Mexicali75, Hoggar, and Sooty have the best ranking with highest grain yield. The mean values over years of Sa% varied between 47.91% for the genotype Oued Zenati and 59.45% for Waha. The genotypes with highest values for the parameter mid-senescence (Σ_{50s}) are the most tolerant and adapted genotypes.

Keywords: durum wheat, senescence, screening, semi-arid

1. Introduction

Durum wheat is one of the most cultivated cereals in the world; it is growing under the Mediterranean regions [1]. Water stress is the abiotic stresses limiting wheat distribution and productivity [2]. Water stress adaptation is considered as the major aim for breeding target in the stabilization of crop performance, by breeders and molecular biologists; at the moment, there is a lack of information to be able to measure with precision the plant resistance under drought stress conditions [3]. Photosynthesis is the primary source of dry biomass production and grain yield in plants. The improvements of leaf photosynthesis have occurred with the advance of breeding high-yielding cultivars. During the period of wheat spike growth, the important moment of assimilation that supplies carbon for the grain depends on the amount and quality of light on the surface of the green area after anthesis. This assimilation area normally decreases due to natural senescence and various stresses. Senescence is considered the final stage in leaf development; senescence in plants is defined as the age-dependent programmed degradation and degeneration process of cells, organs, or the entire organism, leading to death [4]. The most remarkable events in leaf senescence are the loss of chlorophyll and the disassembly of the photosynthetic apparatus, which result in decreases in the photosynthetic energy conversion capacity and efficiency. In addition, chloroplasts of senescing leaves show reduced volume, their shape is spherical, and the thylakoid

 IntechOpen

system is reduced. In cereals, the processes involved in senescence are important because they occur during grain filling, and evidence suggests that early senescence may be yield-limiting [5]. Wheat genotypes vary in the timing of senescence initiation and also in the subsequent rate of leaf senescence. In wheat, the senescence rate was also found to be related to the yield under drought conditions [6, 7]. The quest of the causes of differences in leaf photosynthetic rate among interspecies and/or intraspecies of crops may be one of the important strategies of crop engineering [8]. In all these studies, leaf senescence was evaluated visually. Since senescence corresponds to yellowing due to chlorophyll loss [5], the identification of senescent parts of the leaf is quite easy. In this work, we used an alternative method for the evaluation of the leaf senescence based on numerical analysis of image. In addition, we study the efficiency of using the flag leaf senescence as tools for select adapted durum wheat genotypes under semi-arid conditions.

2. Materials and methods

2.1 Plant material and growth conditions

A set of 10 genotypes of durum wheat (*Triticum durum* Desf.) (**Table 1**) were planted during four cropping seasons (2009–2013), in the experimental fields of INRAA, Setif, Algeria (5°20′E, 36°8'N, 958 m above sea level) genotypes were grown in randomized block design with four replicates. Plots were 5 m × 6 rows with 0.20 m row spacing, and sowing density was adjusted to 300 g m^{-2}.

2.2 Agronomical and physiological measurements

Grain yield (GY) is determined from sub-samples taken from harvested grains of each plot. Leaf senescence (S) was evaluated by numerical image analysis (NIA) according to Hafsi et al. [9]. Leaves were photographed on black surface, between 11:00 and 12:00 solar time with a color digital camera (Canon, Power Shot A460, AiAF, China). Images were analyzed using IPP (Image Pro Plus, Version 4, Media Cybernetics, Silver Spring, MA, USA) software. Senescence was expressed as the ratio of senesced area to total leaf area (in %). Measurements were carried out 10 times between flowering and the end of senescence on three flag leaves for each genotype. Ten dates of assessments were expressed in sums of temperatures after flowering ($\Sigma t_1 - \Sigma t_{10}$) and the corresponding senescence values ($S_1 - S_{10}$). In addition, the date of mid-senescence (Σ_{50}) was evaluated from the experimental curves S = f (Σ_t) as the sum of temperature corresponding to an S value of 50%. Data were analyzed using Costat; the analysis of variance was performed for senescence parameters and grain yield. Linear correlation analysis was used to determine the relationships between the traits measured.

Genotype	Origin	Genotype	Origin
Bousselem	ICARDA/CIMMYT	Altar84	CIMMYT
Hoggar	Spain	Dukem	CIMMYT
Oued Zenati	Algeria	Kucuk	CIMMYT
Polonicum	Algeria	Mexicali75	CIMMYT
Waha	ICARDA/CIMMYT	Sooty	CIMMYT

Table 1.
Name and origin of tested genotypes.

3. Results and discussion

The ANOVA analysis demonstrates significant effect of genotypes and years on senescence parameters and GY. Based on the means comparison, the values of mean grain yield (2009–2013) varied from 37.84 q/ha for Oued Zenati to 44.7 q/ha for

Genotype	Grain yield (q/ha)				Mean over
	2009/2010	2010/2011	2011/2012	2012/2013	all seasons
Oued Zenati	25.50(ab)	52.20(d)	21.45 (b)	47.11(ab)	37.84(b)
Altar84	29.31(a)	55.94(bcd)	24.86 (ab)	64.97(a)	44.79(a)
Sooty	26.56(ab)	63.14(abc)	27.33 (ab)	52.92(ab)	44.29(ab)
Polonicum	24.68(ab)	56.47(abcd)	32.68 (ab)	55(ab)	43.30(ab)
Waha	26.93(ab)	64.63(a)	35.24 (a)	37.31(b)	43.18(ab)
Dukem	22.00(b)	63.94(ab)	29.75 (ab)	44.44(ab)	41.87(ab)
Mexicali 75	31.93(a)	59.64(abcd)	32.90 (ab)	49.34(ab)	44.69(a)
Kucuk	26.50(ab)	53.96(d)	36.87 (a)	47.87(ab)	42.54(ab)
Hoggar	29.68(a)	60.05(abcd)	30.23 (ab)	47.03(ab)	43.42(ab)
Bousselem	29.81(a)	55.01(cd)	36.87 (a)	37(b)	41.26(ab)
Mean	28.00(c)	59.04(a)	30.81(c)	48.3(b)	42.72
Min	22.00	52.2	21.45	37.00	37.84
Max	31.93	64.63	36.87	64.97	44.79
Genotype effect	***	***	***	***	***
LSD 5%	6.45	8.15	13.60	22.26	6.45
Year effect		***			
LSD 5%		4.37			

N.B: Means followed by the same letter are not significantly different ($P \leq 0.05$).

Table 2.
ANOVA analysis and means comparison of grain yield over four cropping seasons.

Genotype	Ranking based on GY				Mean ranking	SD of ranking
	2009/2010	2010/2011	2011/2012	2012/2013		
Oued Zenati	8	10	9	6	8	1.48
Altar84	4	7	8	1	4	2.74
Sooty	6	3	7	3	3	1.79
Polonicum	9	6	4	2	5	2.59
Waha	5	1	2	9	2	3.11
Dukem	10	2	6	8	7	2.96
Mexicali 75	1	5	3	4	1	1.48
Kucuk	7	9	1	5	6	2.96
Hoggar	3	4	5	7	3	1.48
Bousselem	2	8	1	10	5	3.83

Table 3.
Ranking of tested genotypes based on the grain yield.

Altar$_{84}$ with general mean of 42.71 q/ha. Based on the climatic data, the defavorable cropping season is the first one (2009–2010) with mean grain yield equal 27.29 q/ha; during this season, the grain yield varied between 22.0 q/ha for Dukem to 31.93 q/ha for Mexicali$_{75}$. In addition, the best season is 2010–2011 with mean grain yield of 58.49 q/ha, the highest grain yield registered by the genotype Waha (64.63 q/ha) (**Table 2**). The ranking based on the mean grain yield demonstrates that the genotypes Mexicali$_{75}$, Hoggar, and Sooty (**Table 3**) have the best ranking with low values of standard deviation in the changement of ranking over years (1.48, 1.48, and 1.79, respectively); the mean grain yield of these genotypes varied between 44.69, 44.29, and 43.42 q/ha, respectively. A highly significant genotype and years effects was noted for Sa% (average senescence) and the date of mid-senescence (Σ_{50s}) (**Table 4**); the mean values over years of Sa% varied between 47.91% for the genotype Oued Zenati and 59.45% for Waha. For the last parameter

Genotype	2009/2010		2010/2011		2011/2012		2012/2013		Mean over all seasons	
	S_a %	Σ_{50S}	Sa %	Σ_{50S}	Sa %	Σ_{50S}	Sa %	Σ_{50S}	Sa %	Σ_{50S}
Oued Zenati	49.30 (a)	290.9 (d)	44.51 (d)	356.78 (f)	48.56 (e)	350.01 (g)	49.26 (e)	240.95 (cd)	47.91 (e)	309.66 (g)
Altar$_{84}$	38.96 (e)	333.54 (a)	58.94 (ab)	593.49 (cb)	63.26 (ab)	596.72 (bc)	49.9 (e)	283.68 (ab)	52.77 (bc)	451.86 (b)
Sooty	42.57 (cd)	305.17 (c)	55.5 (bc)	594.93 (cb)	56.69 (cd)	598.16 (b)	56.82 (b)	196.90 (ef)	52.89 (bc)	423.79 (e)
Polonicum	43.24 (c)	312.82 (b)	51.53 (c)	479.54 (e)	55.85 (cd)	489.44 (e)	51.25 (d)	247.20 (c)	50.47 (d)	382.25 (f)
Waha	48.07 (a)	269.77 (e)	63.44 (a)	578.72 (c)	67.76 (a)	584.29 (c)	58.51 (a)	239.23 (cd)	59.45 (a)	418.00 (e)
Dukem	40.31 (e)	298.59 (c)	60.26 (ab)	515.37 (d)	57.64 (cd)	518.60 (d)	53.68 (c)	217.24 (de)	52.97 (bc)	387.45 (f)
Mexicali 75	35.31 (f)	338.85 (a)	54.18 (bc)	612.43 (ab)	55.63 (d)	615.66 (a)	47.35 (f)	289.63 (a)	48.12 (e)	464.14 (a)
Kucuk	45.19 (b)	286.63 (d)	54.12 (bc)	625.25 (a)	60.83 (bcd)	628.48 (a)	57.20 (ab)	190.54 (f)	54.33 (b)	432.73 (d)
Hoggar	40.95 (de)	316.92 (b)	57.53 (bc)	594.25 (cb)	60.12 (bcd)	597.48 (bc)	49.36 (e)	260.77 (bc)	51.99 (cd)	442.35 (c)
Bousselem	42.8 (cd)	334.46 (a)	56.79 (bc)	470.64 (e)	61.11 (bc)	445.87 (f)	47.84 (f)	281.84 (ab)	52.13 (cd)	383.20 (f)
Mean	42.67 (d)	308.76 (b)	55.68 (b)	542.14 (a)	58.74 (a)	542.47 (a)	52.12 (c)	244.80 (c)	52.3	409.54
Min	35.31	269.77	44.51	356.78	48.56	350.01	47.35	190.54	47.91	309.66
Max	49.3	338.85	63.44	625.25	67.76	628.48	58.51	289.63	59.45	464.14
Genotype effect	***		***		***		***		***	
LSD 5%	1.24	5.34	4.12	16.23	5.28	13.33	1.33	26	1.68	8.16
Years effect	***		***		***		***		***	
LSD 5%	1.01	5.20	1.01	5.20	1.01	5.20	1.01	5.20		

*** *$P < 0.001$*

Table 4.
ANOVA analysis and means comparison of senescence parameters over four cropping seasons.

Genotype	Ranking based on Senescence parameters								Mean ranking		Total mean ranking	SD of ranking	
	2009/2010		2010/2011		2011/2012		2012/2013		Sa%	Σ_{50S}		Sa%	Σ_{50S}
	Sa%	Σ_{50S}	Sa%	Σ_{50S}	Sa%	Σ_{50S}	Sa%	Σ_{50S}					
Oued Zenati	10	8	1	10	1	10	3	9	2	10	5	3.49	0.80
Altar84	2	3	8	5	9	5	5	7	7	4	4	2.45	1.74
Sooty	5	6	5	3	4	3	8	10	5	5	3	1.47	2.58
Polonicum	7	5	2	8	3	8	6	8	3	9	5	1.94	1.36
Waha	9	10	9	6	10	6	10	3	10	8	8	0.49	2.23
Dukem	3	7	10	7	5	7	7	1	8	6	7	2.42	2.40
Mexicali75	1	1	4	2	2	2	1	6	1	1	1	1.10	1.85
Kucuk	8	9	3	1	7	1	9	4	9	2	4	2.23	2.93
Hoggar	4	4	7	4	6	4	4	5	4	3	2	1.26	0.63
Bousselem	6	2	6	9	8	9	2	2	6	7	6	1.96	3.29

Table 5.
Ranking of tested genotypes based on the senescence parameters.

(Sa%), the genotypes with lowest values are the preferable and adapted genotype. However, the genotypes with highest values for the parameter mid-senescence (Σ_{50s}) are the most tolerant and adapted genotypes; the mean values over years of mid-senescence varied between 464.14°C for the genotype Mexicali75 and 309.66°C for the genotype Oued Zenati. The total mean rankings based on the senescence parameters demonstrate that the genotypes Mexicali75, Hoggar, and Sooty are the best genotypes under these conditions (**Table 5**). Our study showed significant correlation between grain yield and the parameter mid-senescence (Σ_{50s}) (r = 0.91*). Over 50 years ago, it was realized that the diversity in yield for most crops is mainly a consequence of variation in the duration, rather than the rate of photosynthetic activity [10], and so, delayed leaf senescence (i.e., stay-green) has long been considered to be a desirable trait in cereal breeding. Total flag leaf photosynthesis, chlorophyll content, the onset of senescence (at low nitrogen availability), and green leaf duration have all been found to be positively correlated with wheat grain yield [11].

4. Conclusion

The results of this study demonstrate that the genotypes with highest values for the parameter mid-senescence (Σ_{50s}) are the most tolerant and adapted genotypes. Based on the mean grain yield ranking, the genotypes Mexicali75, Hoggar, and Sooty have the best grain yield. In addition, the screening based on the senescence parameters showed that the genotypes Mexicali75, Hoggar, and Sooty are the preferable and adapted genotype. The combination between the rankings based on the GY and senescence parameters demonstrate that the genotypes Mexicali75, Hoggar, and Sooty are the best and recommended genotypes under this condition.

Author details

Hafsi Miloud[1*] and Guendouz Ali[2]

1 Laboratory for the Improvement and Development of Plant and Animal Production, Department of Agronomy, Ferhat ABBAS University of Setif, Algeria

2 Research Unit of Setif INRAA, National Institute of Agronomic Research of Algeria, Algeria

*Address all correspondence to: hafsimiloud@gmail.com

References

[1] Baldy C. Comportement des blés dans les climats méditerranéens. Ecologia Mediterranea. 1986;**XII**(3–4): 73-88

[2] Mastrangelo AM, Rscio A, Mazzucco L, Russo M, Cattivelli L, Di Fonzo N. Molecular aspects of abiotic stress resistance in durum wheat. Options Méditerranéennes, Series A. 2000;**40**:207-213

[3] Blum A. Crop response to drought and the interpretation of adaptation. Plant Growth Regulation. 1996;**20**: 135-148

[4] Lim PO, Kim HJ, Nam HG. Leaf senescence. Annual Review of Plant Biology. 2007;**58**:115-136

[5] Patterson TG, Moss DN. Senescence in field –grown wheat. Crop Science. 1979;**19**:635-640

[6] Mi GH, Tang L, Zhang FS. Nitrogen uptake and translocation during grain formation of two wheat cultivars with contrasting maturity appearance. Journal of South China Agricultural University. 1999;**3**:453-457

[7] Pajević S, Krstih B, Stankovih Z, Plešnicar M, Dencih S. Photosynthesis of flag leaf and second wheat leaves during senescence. Cereal Research Communications. 1999;**27**:155-162

[8] Jiang H, Wang XH, Deng QY, Yuan LP, Xu DQ. Comparison of some photosynthetic characters between two hybrid rice combinations differing in yield potential. Photosynthetica. 2002; **40**:133-137

[9] Hafsi M, Mechmeche W, Bouamama L, Djekoune A, Zaharieva M, Monneveux P. Flag leaf senescence, as evaluated by numerical image analysis, and its relationship with yield under drought in durum wheat. Journal of Agronomy and Crop Science. 2000;**185**:275-280

[10] Watson DJ. The physiological basis of variation. Advances in Agronomy. 1952;**4**:101-145

[11] Wang H, Mc Caig TN, De Pauw RM, Clarke JM. Flag leaf physiological traits in two high-yielding Canada Western red spring wheat cultivars. Canadian Journal of Plant Science. 2008;**88**:35-42

Chapter 7

Ethiopian Common Medicinal Plants: Their Parts and Uses in Traditional Medicine - Ecology and Quality Control

Admasu Moges and Yohannes Moges

Abstract

The main purpose of this review is to document medicinal plants used for traditional treatments with their parts, use, ecology, and quality control. Accordingly, 80 medicinal plant species were reviewed; leaves and roots are the main parts of the plants used for preparation of traditional medicines. The local practitioners provided various traditional medications to their patients' diseases such as stomachaches, asthma, dysentery, malaria, evil eyes, cancer, skin diseases, and headaches. The uses of medicinal plants for human and animal treatments are practiced from time immemorial. Stream/riverbanks, cultivated lands, disturbed sites, bushlands, forested areas and their margins, woodlands, grasslands, and home gardens are major habitats of medicinal plants. Generally, medicinal plants used for traditional medicine play a significant role in the healthcare of the majority of the people in Ethiopia. The major threats to medicinal plants are habitat destruction, urbanization, agricultural expansion, investment, road construction, and deforestation. Because of these, medicinal plants are being declined and lost with their habitats. Community- and research-based conservation mechanisms could be an appropriate approach for mitigating the problems pertinent to the loss of medicinal plants and their habitats and for documenting medicinal plants. Chromatography; electrophoretic, macroscopic, and microscopic techniques; and pharmaceutical practice are mainly used for quality control of herbal medicines.

Keywords: medicinal plants, herbal medicine, chromatography, histological techniques, pharmaceutical practices, microscopic and macroscopic examination

1. Introduction

Medicinal plants are very vital in their uses for medication, besides providing ecological, economic, and cultural services. The world primary means of treating diseases and fighting infections have been based on the use of medicinal plants. From ancient times, plants have been rich sources of effective and safe medicines [1]. Globally, about 64% of the total world population is reliant on traditional medicine for their healthcare needs [2]. According to the World Health Organization (WHO), nearly 3.5 billion people in developing countries including Ethiopia believe in the efficiency of plant remedies and use them regularly [3].

IntechOpen

Ethiopia is located in the Horn of Africa between 3 and 15° northing, latitude, and 33 and 48° easting, longitude, and is also comprised of nine national regional states and two administrative states with varied agroecological zones. Since the country is characterized by a wide range of ecological, edaphic, and climatic condition, Ethiopia is also very diverse in its flora composition [4]. The flora of Ethiopia is estimated to contain close to 6500–7000 species including medicinal plants; of those, 12–19% are endemic to the country [5]. The medicinal plants have been used for various types of human and animal treatments in the country. According to [6, 7], in Ethiopia, about 80% of human population and 90% of livestock rely on traditional medicine. As also stated by many authors (e.g. [6, 7]), the medicinal plants have shown very effective medicinal values for some diseases of humans and livestock.

Even due to the trust of communities on medicinal values of traditional medicines, culturally associated traditions, and their relatively low cost, medicinal plants are highly demanded in Ethiopia [7]. Inadequate health centers and shortage of medicines and personnel in clinics might be the other reasons for driving the people of Ethiopia, in general, and the low-income community and the rural people, in particular, to the traditional health centers, whereby increasing the demand of medicinal plants.

However, these plants have got little attention regarding the documentation of scientific names, uses, ecology, and conservation in Ethiopia, in particular and world-wise, in general. Moreover, in Ethiopia, traditional medicine is faced with a problem of sustainability and continuity mainly due to the loss of taxa of medicinal plants [8, 9] besides having lack of quality control for herbal medicines. The main causes for the loss and decline of diversity of plants in Ethiopia are human-made factors [10–12]. Habitat destruction and deforestation for commercial timber and forest encroachment for urbanization, investment, agriculture, and other land uses are the major causes of the loss of many thousand hectares of forest that harbor medicinal plants yearly for the past several decades. In addition to these, the medicinal plant materials and associated traditional knowledge are being lost due to the lack of systematic conservation, research, proper utilization, and documentation [13]. The knowledge on identifying and managing the medicinal plants with their parts, use, and ecology is mostly associated with local and elder people, who transmitted their knowledge verbally. Such verbal transmissions of knowledge on medicinal plants have thus resulted in eroding and loss of knowledge and the plant materials as well. The quantity and quality of the safety and efficacy data on traditional medicine are also far from sufficient to meet the criteria needed to support its use worldwide [14]. Therefore, assessing and documenting the medicinal plants along with their useful medicinal parts, use, and ecology in Ethiopia, as well as revising the quality control for herbal materials and medicine, are very crucial for giving priority to their conservation and sustainable utilization.

2. Materials and methods

The materials for this review were published documents. However, regarding the screening of medicinal plants, some medicinal plants not yet identified or available in more than one article being revised during this revision time, and published before 2000 with their uses, were not listed and included for this review analysis so as to increase the quality of the present review, provide the current information to the readers, and restrict the revised papers. Based on this, of the total (32) revised documents, 15 articles, which are assessing the different medicinal plants with their uses and parts, were revised for documenting the medicinal plants for this review.

Additionally, the habitats (ecology) of each medicinal plant were assessed from the Flora Volumes of Ethiopia and Eritrea and [15], besides the articles revised for listing the medicinal plants for this review. The data were analyzed and described quantitatively using frequency, percentage, tables, and figures via applying Microsoft Excel Spreadsheet 2010 and SPSS with version 20, as well as qualitatively using content analysis, narrating via drawing sub-contents.

3. Medicinal plants: their parts, uses, and ecology reviewed

Traditional healers in Ethiopia utilize the herbal resources available in nature for various disease treatments. As reported before, approximately 800 species of the medicinal plants grown in Ethiopia are used for treating about 300 medical conditions [16]. However, based on the present review, the number of medicinal plants and the treatments/medications identified and listed are limited as presented here under section by section.

3.1 Medicinal plants and their growth forms and parts used

3.1.1 Composition and growth forms of medicinal plants

As reported by many authors [6, 7, 12, 13, 17–27], there are different types of medicinal plant species with their parts, habitats, and disease types being treated and described here in **Table 1**. Accordingly, as depicted in **Table 1**, there were 80 medicinal plant species with 63 genera, used by the local communities for various human treatments. Among other revised, the common medicinal plants used for treating and curing various diseases are *Aloe* species, *Eucalyptus globulus*, *Hagenia abyssinica*, *Cupressus macrocarpa*, *Buddleja polystachya*, *Acmella caulirhiza*, *Acacia* species, *Citrus* species, *Clematis* species, *Coffee Arabica*, *Croton macrostachyus*, *Euphorbia* species, *Ficus sycomorus*, and *Moringa stenopetala* (**Table 1**).

Based on the review, all plant growth forms were not equally used as remedies, because of the difference in distribution among the growth forms. Accordingly, the life forms of medicinal plants reviewed constituted 18 trees (22.78%), 23 shrubs (29.11), 29 herbs (36.71%), 3 climbers (3.81%), 4 trees/shrubs (5.06%), and 2 herbs/shrubs (2.53%) (**Figure 1**). Of all life forms, herbs were, thus, the major medicinal plants used by the community for human treatment followed by shrubs and trees.

3.1.2 Medicinal plant parts used for preparation of traditional remedies

The review indicated that the plant parts used for medication preparation by the traditional healers are variables. Healers mostly used fresh specimens from commonly available plants [25] to prepare remedies for their patients; this might be mostly due to the effectiveness of fresh medicinal plant parts in treatment since the contents are not lost before use compared to the dried ones [12]. As also referred from many authors, the traditional healers have harvested leaves, roots, barks, seeds, fruits, stems, flowers, barks, seeds, or latex of medicinal plants (**Figure 2**) to prepare their traditional medicines for their patient treatments. As depicted in **Figure 2**, most remedies were prepared from the leaf (32.98%) and root (29.79%) parts of the medicinal plants to treat the diseases compared to the other parts of them. This finding of the review is in line with the findings of the majority of authors' papers (e.g. [18, 25, 27]). The main reason that many traditional medicine practitioners used the leaf parts compared to others for remedial preparation is due to their accessibility and for preventing them from extinction [25]. In fact,

Scientific names	Local name	Ha.	Habitat	Parts used	Uses [references cited]
Acacia abyssinica Hochst ex. Benth	Qontir	S	Deciduous bushland	Leaves	Used for treating goiter [18, 22]
Acacia nilotica (L.) Del.	Girar	T	Dry bushland	Fruits Leaflets	For treating diarrhea, diabetes, sore gum, hemorrhage, and loose teeth For curing sickness of stomach [19, 21, 27]
Acacia albida Del.	Grar	T	Dry bushland	Latex	Latex from the stem pounded is taken with honey for curing amebiasis; for treating fire wound [13, 27]
Acmella caulirhiza Del.	Yemdir berbere	H	Wetlands, forest floors, stream banks	Leaves Flowers	Used for curing tonsillitis via chewing the flowers and spitted on tonsillitis [18, 22]
Aerva javanica (Burm.f.) Schultes	Nech shinkur	S	Dry sandy plains, dried river course	Root	For treating cancer [20, 24]
Allium sativum L.		H	Irrigable cultivated land, home garden	bulb	For preventing and treating malaria [7, 13, 18, 22, 23, 25]
Amaranthus caudatus L.	Chigogot	H	Roadsides, riverbanks, floodplain	Leaves	Used for curing diarrhea via pounded and boiled leaves [18, 22]
Aloe monticola Reynolds	Eret	H	Steep bare mountain slopes	Root	For also curing anthrax by pounding the root and mixing it with cold water and local alcohol [12, 22]
Aloe macrocarpa Reynolds		H	Rocky slopes	Leaves	For preventing wart by powdering leaf and then mix it with honey [12, 22, 26]
Artemisia abyssinica Sch. Bip. ex. Rich	Chigugn	H	Juniper forest, open grassland, fallow fields	Fresh root	For preventing evil spirit by smelling and drinking after crushing the root and normalizing it in water [7, 22, 25]
Asparagus africanus L.	Yeset qest	H	Acacia woodland Forest margins	Roots	For curing uterine and breast cancer [17, 20, 24]

Scientific names	Local name	Ha.	Habitat	Parts used	Uses [references cited]
Barleria eranthemoides R. Br. ex C. B. Cl	Yeset af	S	Acacia woodland Scrublands	Roots	For curing hear burn [12]
Bersama abyssinica Fresen.	Azamir	T	Riverine forest, rainforest	Leaves-stem	For treating wound by squeezing the leaves and creaming on the wound [22, 24]
Bridelia scleroneura Mul. Arg.		T	Open woodland Dry riverine forest	Seeds	For curing skin diseases by crushing and applying on wound parts [12, 18, 19]
Brucea anti dysenterica Fresen.	Abalo	S/T	Montane, evergreen forest margins	Leaves	For treating cancer, skin problem, leprosy, and external parasites [6, 25]
Buddleja polystachya Fresen.	Anfar	T	Degraded woodland in cultivated fields, around houses	Leaves	For treating the cattle eye diseases by chewing and spitting on the affected area [18, 22]
Calpurnia aurea (Ait.) Benth.	Digita	S	Forest margins, bushland/grassland, favored by over grazing	Leaves Roots Seeds	For preventing poisonous snake bite by boiling the leaves and drinking with honey [12, 24] For curing amebiasis by crushing and boiling with leaf of coffee for drink. The seeds can be used as a fish-poison or as a cure for dysentery [12]
Capparis tomentosa Lam		S	Riverine forest, grassland with scattered trees	Bark	For curing sore, anthrax, and evil eye using the powder of the bark with hot water [18, 20]
Carica papaya L.	papaya	T	Home gardens, small and large plantations	Seeds	Used for treating diarrhea and ascariasis by drinking the ground and boiled seeds with honey [12, 19, 27]
Carissa edulis (Forsk.)	Agam	S	Open *Acacia* bushland	Root	Used for shorten the labor period just before delivery of women [19, 21]

Scientific names	Local name	Ha.	Habitat	Parts used	Uses [references cited]
Carissa spinarum L.	Agam	S	Disturbed areas, along edges of roads, riverine vegetation	Roots	Used for preventing evil eye by inhaling the smoke of pounded roots. It is also used for treating wounds via applying the powder of the roots [12, 17, 19, 27]
Clausena anisata (Wild.) Benth.	Limich	S	Montane forest margins, moist forest, secondary bushland	Leaves	For treating skin irritation by pounding together the leaf of *C. anisata*, *Solanecio gigas*, and *Justicia schimperiana* [6, 18, 20, 22]
Citrus aurantifolia Swingle	Bahre-Lomi	T	In lowlands, evergreen forest	Fruit	For treating dermatophyte [6, 12, 19]
Citrus sinensis (L.) Engl.	Birtukan	S	Cultivated in irrigable areas	Fruit Bark	Used for treating stomach infection and wound [12, 18]
Clematis hirsuta Per.	Nech Azo hareg	Cl	Edges and remnants of montane forest, roadsides, paths	Leaves/ stems Barks	Used for treating tumor/cancer on the neck [19, 24]
Clematis simensis Fresen.	Hareg	Cl	>>	Leaves Root	Used for curing wound and stomachache [12, 18]
Clerodendrum myricoides (Hochst.)	Misrich	S	Not specified yet	Root	Used for treating earache and headache [12, 20]
Coffee arabica L.	Buna	S	In shaded coffee plantations	Seeds	For curing diarrhea by pounding and mixing with honey [6, 12, 18]
Cordia africana Lam.	Wanza	T	Moist evergreen forest, riverine vegetation, woodland, grassland	Roots	For curing itching via applying the powder of the root on the area [6, 12, 13, 18, 19]
Crinum abyssinicum	Yejb shinkurt	H	Waterlogged valley grasslands, swampy or along stream banks, fallow fields	Leaves	Used as treatment of tumor in general [13, 20, 24, 25]
Croton macrostachyus Hochst. ex Del.	Bisana	T	Forest margin, edges of roads, disturbed areas, woodland	Bark	For curing splenomegaly and gonorrhea [12, 17, 18, 20, 22, 25]
Croton zambesicus	Bisana	T	Stony streambeds, within broad-leaved deciduous woodland	Bark	Used for treating mental disturbance [21, 27]

Scientific names	Local name	Ha.	Habitat	Parts used	Uses [references cited]
Cucurbita pepo L.	Duba/ Yebarqil	H	Cultivated in home garden, farmland	Leaves	Used as a means of treating gastritis [12, 22]
Datura stramonium L.	Atse-faris	H	Disturbed places, waste ground, near water holes, roadsides	Seed	Used for treating depression [22, 25]
Dodonaea angustifolia (L.fil.) J.G.West		S	Not defined	Root	For curing toothache and wound [6, 7, 12, 18, 23]
Dorstenia barnimiana Schwienf.	Worq-bemeda	H	Woodland bushland, upland grassland, evergreen bushland	Roots/ tubers	For treatment of tumor visible in body surface [20, 24]
Echinops kebericho, Mesfin	Kebericho	H/S	Montane *Acacia* woodland, disturbed bushland	Root	For treating toothache, vomiting, and headache [22, 27]
Ehretia cymosa Thonn.	Oulaga	H/S	Montane and riverine forest, evergreen bushland, hedgerows around compounds	Leaves	Used for curing bleeding, fibril illness [12, 18]
Eucalyptus globules Labill.	Nech-bahirzaf	T	A wide variety sites (plantations)	Leaves	Used for treating influenza and allergic [7, 13, 18, 22, 23, 26]
Euclea racemosa L.	Dedeho	S	Open montane and bushland; in clearings and along margins	Roots	For treating evil spirit, evil eye, and heartburn [12, 17]
Euphorbia tirucalli L.	Qinchib	S	Live fence of home garden	Roots Latex	Used as treatment of tumor/cancer [7, 12, 23]
Euphorbia abyssinica J. F. Gmel.	Qulkual	T	Steep rocky hillsides, around churches; live fence at higher altitudes	Latex	For treating skin cancer [20, 22]
Rhus natalensis Beru ex Krauss.		H	*Acacia-Commiphora* woodland, wooded grassland, near rivers on various soil types	Leaves	Used for treating skin wound and boils [12, 21]
Ficus sycomorus L.	Banba	T	River and lake margins, woodland, forest edges and clearings, wooded grassland	Bark	For curing hepatitis [18, 19, 22]
Gladiolus schweinfurthii (Baker) Goldblatt and M.P. de Vos	Milas golgul	H	Open grassland; *Acacia* woodland; rocky limestone slope	Root	Used for treating headache [12, 22, 24]

Scientific names	Local name	Ha.	Habitat	Parts used	Uses [references cited]
Glinus lotoides L.	Meterie/ Amkin	H	Disturbed sites	Leafy stem	For treating tapeworm
Guizotia scabra (Vis.) Chiov.	Mechi	H	Open wasteland, grassland, weed of cultivation, roadside ditches, riverbanks	Leaves	Used as wound treatment [6, 22]
Justicia schimperiana Hochst. ex A. (Nees) T. Anders	Sensel	S	Open woodland, riverine vegetation, live fence of house	Leaves	For preventing bat urine [6, 7, 12, 18, 20, 26]
Harrisonia abyssinica	Ddugot	S	Montane forest and grassland	Barks	For giving human physical strength [21]
Hagenia abyssinica (Bruccie) T.F.Gmel	Kosso	T	Montane forest and grassland Moist evergreen forest	Fruits	Tapeworm [7, 23, 25, 26]
Laggera crispata (Vahl.)	Gemie	S	Cultivation and waste places, grassland, riverbanks	Leaves	For preventing dizziness [12, 20]
Maesa lanceolata Forssk		T/S	Gallery forest, margin of evergreen forest, along river banks and streams, open woodland and valleys	Bark	For curing elephantiasis [6, 18, 26]
Malva verticillata L.	Lut	H	Paths and clearings in upland forest, upland grassland, cultivated areas near houses	Root	For curing cancer/ tumor [6, 18, 24, 25]
Mimusops kummel A. DC.	Safa/kummel	T/S	In gullies, in riverine forest, in riparian woodland, in woody vegetation on lake shores	Root	Used for preventing lung cancer [12, 18]
Moringa stenopetala (E.G. Baker) Cufod.	Shiferaw	T	Cultivated in terraced fields, gardens, small towns, in riverine and woodland	Root	Used for asthma relief [7, 12, 21]
Musa sapientum L.	Koba	H	Cultivated on large irrigated farms and in house gardens	Bulb	It is taken as an abortion medicine [19, 21]
Nicotiana tabacum L.	Timbaho	H	Cultivated in villages, home gardens, tobacco farms	Leaves	For treating snakebite [6, 12, 18]
Nigella sativa L.	Tikur azmud	H	Cultivated in homesteads, in fields; growing in wild	Seed	Used as treatment of headache [18, 22]

Scientific names	Local name	Ha.	Habitat	Parts used	Uses [references cited]
Ocimum lamiifolium Hochst. ex. Benth.	Damakesie	S	*Acacia-Commiphora* bush- and woodland, limestone slopes, home gardens	Leaves	Fibril illness [7, 12, 18, 20, 22]
Olea europaea L.	Woira	T	Home garden, monasteries and churches, woody vegetation	Leaves/ roots	For curing dysentery, wound stomachache, bone TB [6, 12, 17, 18, 20, 26]
Opuntia ficus-indica (L.) Miller	Yebereha qulkual	S	Disturbed areas, degraded areas, live fence of houses	Leaves	For killing malaria vectors [22, 25]
Plumbago zeylanica L.	Amira	H	Disturbed habitats by roads and paths, bushland, woodland, savannah	Root	For preventing gonorrhea and hemorrhoids as well as for toothache [12, 20, 22]
Verbascum sinaiticum Benth.		H	Disturbed sites	Root/ leaves	For treating heart disease, cancer, trypanosomiasis [6, 20, 27]
Premna schimperi Engl.	Chocho	*S*	Degraded and secondary forests, grassy meadows and along paths in forests	Root Leaves	Used for treating mastitis Used for preventing boils [12, 18]
Solanum nigrum	Embuay	H	In cultivation and ruderal areas, on road-, hill-, river- or streamsides; in bushland areas	Leaves roots, stems	Leaf, root, and stalk are used for cancerous sores and wound treatments. Stems eaten as pot herb for virility in men and for dysmenorrhea in females, for dysentery, and sore throat [21, 24]
Solanum incanum L.	Tikur awud	H	Cultivated and riverine gallery forest, disturbed habitats	Leaves/ roots	Used for curing bleeding, menstruation, amebiasis [12, 17–20]
Stephania abyssinica (Dill. and A. Rich.) Walp. (Etse Eyesus, Nech- Hareg)	Yayit hareg	Cl	In thickets bordering forest margins, hillsides, cultivated fields, in clearings	Root	For treating external tumor/ cancer and stomachache [6, 12, 8, 24, 25]
Stereospermum kunthianum Cham.	Arziniya	S/T	Open woodland and savanna, widespread in tropical Africa	Bark	Used for treating kidney via drinking the juice crushed from bark [12, 13, 19]

Scientific names	Local name	Ha.	Habitat	Parts used	Uses [references cited]
Tamarindus indica L.	Humer/Roqa	T	Grassland, woodland *Combretum* bushland, riparian	Fruit	Used for curing stomachache; it is also used for treating bile and intestinal worm using the fruit juice with hot water in the morning before breakfast [12, 19]
Thunbergia ruspolii Lindau	Marte	H	*Combretum-Terminalia* woodland, grassland, wooded grassland, evergreen forest, seasonally waterlogged	Not reported	For curing poisonous snakebite [21]
Thymus capitatus (L.) Link	Tosign	H	Not reported	Leaves	For curing stomach diseases, cough, and asthma [21, 25]
Tragia cordata Michx.	Alebilabet	H	Among open rock bushlands	Root	For treating urinary tract and external parasite [12, 18, 19]
Tribulus terrestris L.	Kurnchit	H	Open and disturbed places, often on sandy soils	Stem Fruit Seed	For curing scabrous skin diseases For congestion, headache, hepatitis, liver, vertigo, stomatitis, kidneys, liver, and vision For treating anemia, hemorrhoid coughs, fluxes, and stomatitis [21]
Urtica pilulifera L.	Sama	H	Unknown	Leaves	For curing sore joints by mixing the plant juice with oil; provide cure for rheumatism and hemorrhage [18, 21]
Vernonia amygdalina Del.	Girawa	S	Bush/woodland, forest habitats, home gardens	Leaves	For preventing headache and intestinal worm and for treating tumor/cancer in general [6, 7, 12, 18, 20, 22, 24, 26, 27]
Xanthium strumarium L.	Deha nikel	H	Wet forest margins, in riverine vegetation by streamside	Leaves	Used for treating dandruff [12, 27]

Scientific names	Local name	Ha.	Habitat	Parts used	Uses [references cited]
Ximenia americana L.	Enkoy	S	*Acacia* woodland, *Acacia-Ballanites*, woodland, *Combretum-Terminalia*, wooded grassland	Fruit Kernel Root	Oil from the fruit kernel is applied to fresh wounds to prevent infections and also used by some people, who have their ears or lips pierced Used for treating stomachache and tonsillitis [6, 12, 19, 20]
Warburgia ugandensis Sprague		T	Transitional montane forest, adjacent woodland	Stem	Used for treating boils and cough [12, 17]
Withania somnifera L. Dunal	Gizawa	S	In cultivations, disturbed places in the highlands, on lake shores, along riverbanks in disturbed places in open woodland	Leaves	Used for treating malaria [12, 13, 17]
Ziziphus spina-christi (L.) Desf	Qurqura	T	Wooded grassland, along dry riverbeds, edges of cultivations and home gardens	Fruits	Used for treatments of stomachache, tonic, for tooth aches, and tumors [21, 13]

NB: Ha, habits; T, tree; S, shrub; H, herbs; Cl, climbers; T/S, shrubs/trees; H/S, herbs/shrubs.

Table 1.
List of reviewed Ethiopian medicinal plants used for various traditional disease treatments with their parts and ecology/habitat.

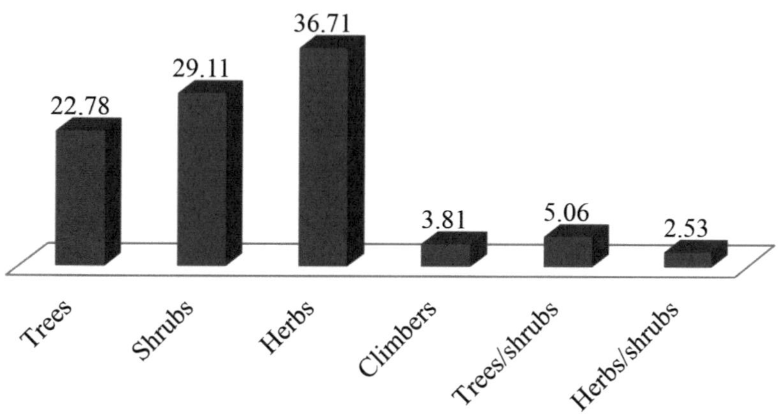

Figure 1.
Life forms/habits of medicinal plants reviewed with their percentage (%).

harvesting the root parts of the medicinal plant for preparation of traditional medicines has negative consequences on the existence of the plants themselves in the future. That is why most of the medicinal plants are currently at risk, declining highly due to them using their root parts besides other human pressures.

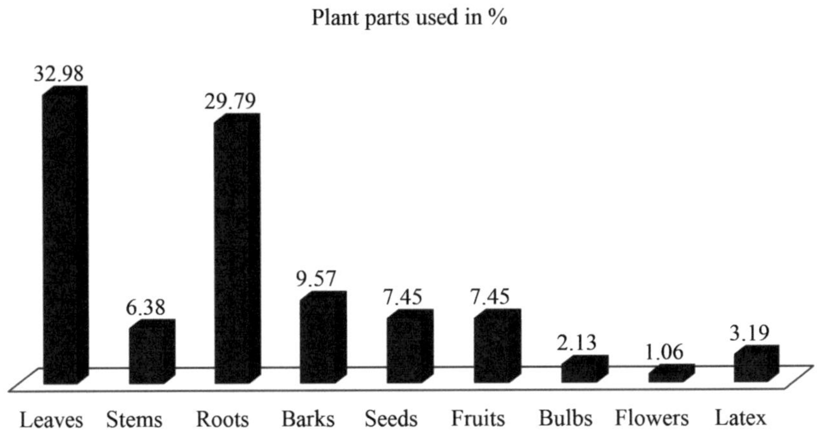

Figure 2.
Distribution of medicinal plant parts used for disease traditional treatments by healers.

3.2 Uses of medicinal plants in treating different disease types

Using these medicinal plants revised in **Table 1**, the local communities could be able to treat about 69 disease types. The disease types treated by these various medicinal plants were skin disease, gonorrhea, diarrhea, wound, tapeworm, snake bites, stomachache, headache, evil eye, heartburn, cancer/tumor, and malaria (see **Table 1** for the detail). Particularly, most of the patients (who come from rural areas) with their perspective disease types have been treated by traditional healers, before coming to clinics and/or hospitals located far away by many kilometers from their residential areas. The disease types most frequently treated by traditional medications (traditional healers) provided by those medicinal plants were stomachaches, wounds, cancers/tumors, skin diseases, headaches, toothaches, and coughs and diarrhea, which took the first, second, third, fourth, fifth, sixth, and seventh ranks, respectively, although the majority of disease types were frequently treated less than four times, ranging from one to three times (**Table 2**). This also points out that one medicinal plant species can be used for treating more than one disease types.

Because of this, medicinal plants are very vital in providing traditional medicines, prepared by local healers, and thereby used for treating and curing different types of diseases that affected the local communities, where they occurred. Even, following the traditional uses and effectiveness of the medicinal plants [23], the traditional healers are also popular by the local societies, providing cultural values. The study of [23] also confirmed that the traditional health practitioners are with a good knowledge of medicinal plants used to treat different diseases of their locals.

In addition to these contributions pertinent to traditional medications and cultural values, the individual medicinal plants could provide regulating, provisioning, and supporting services. For instance, they could provide regulating services via regulating soil erosion, climate change, disease, pollution, and pollination; they also provide provisioning services such as fuel wood, timber for house construction, food (fruits, honey), and fodder and shelter for wild animals [11]. Hence, almost all of the medicinal plants are multipurpose species, providing more than one benefits.

3.3 Ecology and/or habitats of medicinal plants

As referred from the revised documents for this review, the habitat preference of medicinal plants varied from place to place (**Table 1**). As referred in **Table 1** and

No. of disease type	Frequency of treatments	Rank
1 (Stomachaches)	12	1
1 (Wounds)	11	2
1 (Cancer/tumor)	10	3
1 (Skin diseases)	7	4
1 (Headaches)	6	5
1 (Toothaches)	5	6
2 (Cough, diarrhea)	4 (each)	7
8 (Tonsillitis, malaria, evil eye, snakebites, dysentery, boils, throat sore, intestinal worms)	3 (each)	8
10 (Earache, amebiasis, urinary tract, heartburn, external parasites, fibril illness, kidney, liver, hemorrhoids, tapeworms)	2 (each)	9
43 (Elephantiasis, asthma, eye diseases, diabetes, anthrax, leprosy, etc.)	1 (each)	10

Table 2.
*Disease type categories and their rank based on their frequency being treated by different medicinal plant species (as described in **Table 1**).*

Figure 3 drawn from the review, the majority of medicinal plants were available along the edges of river/streams and wetlands, disturbed sites, grasslands, cultivated lands, woodlands, bushland, grasslands, and home gardens. Generally, the majority of medicinal plants were found in wild compared to those plants found in cultivated and home gardens together. Many of the authors of the reviewed articles (e.g. [12, 23, 25]) confirmed that the majority of medicinal plants were collected from natural habitats or wild by traditional practitioners compared from home gardens. Among medicinal plants found along stream/riverbanks (**Figure 3**), the majority of them are supposed to be medicinal plants having herbal life forms/habits (**Figure 1**). This could be due to their shallow roots, which cannot bring water from the deep parts of their habitats.

Because of the anthropogenic factors such as over harvesting, fire/deforestation, agricultural expansion, overgrazing, and urbanization [25, 28], most of the medicinal plants have also been lost. This implies that the availability and accessibility of most medicinal plants in Ethiopia are also very difficult [25]. Hence, most of the medicinal plants were restricted to areas (such as cliffs, hills/mountains, gorges, disturbed areas, riverbanks, and valleys of wild) which are not easily accessible to use/harvest them. Not only is this, but also the knowledge of traditional practitioners pertinent to identification of medicinal plants with their parts and ecology and the process of preparation of herbal medicines and medication with their quality/effectiveness are declined/lost since the knowledge is mostly transferred orally from generation to generation, not documented. Therefore, the effects of human on the natural habitat of medicinal plants are the problems for the conservation of medicinal plants and associated knowledge of traditional healers [12]. With the present ecological and socioeconomic changes, medicinal plants together with the associated ethnobotanical knowledge in Ethiopia are under serious threat and may be lost at alarming rate.

Under such circumstances, the use of plants for medicinal purposes will also decline, and consequently the once effective traditional healthcare system will also be lost [19]. Hence, documenting medicinal plants with their uses and ecology as well as the knowledge of traditional practitioners is so vital. Moreover, it is very essential to give conservation priority for those medicinal plants through

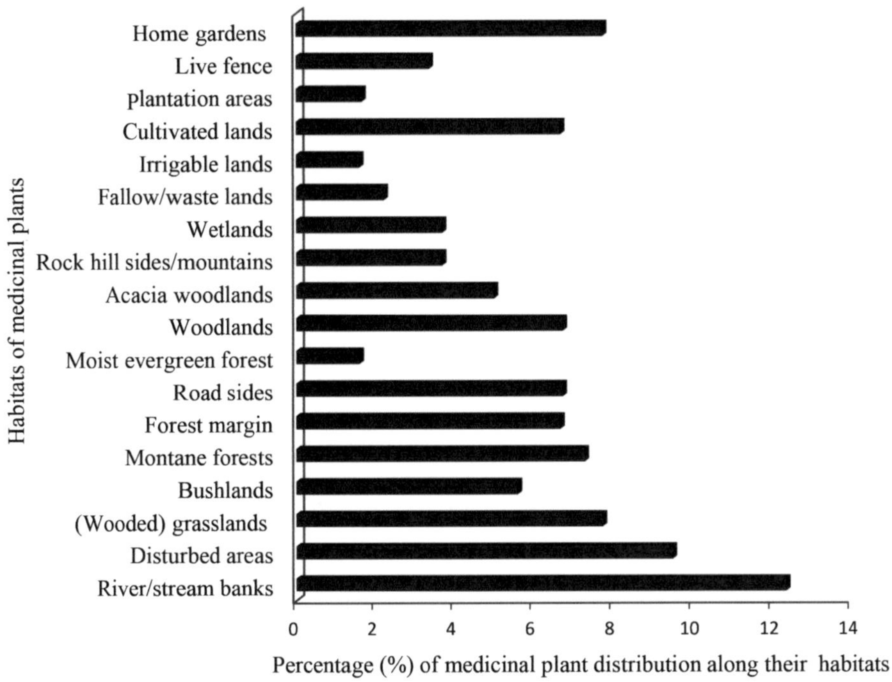

Figure 3.
Summary of distributions of medicinal plants along their major habitat categories in Ethiopia.

protecting them where they are found, propagating them in cultivated areas and home gardens, and creating awareness to the locals. Hence, following community and research-based approach is advised to save medicinal plants from their loss and extinction.

4. Applied plant anatomy: quality control of herbal medicine

4.1 General overview

Plant materials are used throughout developed and developing countries as home remedies, as over-the-counter drug products, and as raw materials for the pharmaceutical industry, which represent a substantial proportion of the global drug market [29]. Thus, the traditional herbal medicines and their preparations have been widely used for thousands of years in many countries. Therefore, it is so essential to overview here some modern control histological techniques or tests, suitable standards, and practical experiences used for assessing the quality of medicinal materials and their products. Quality control of herbal medicine using histological techniques and pharmaceutical practices is also very vital for avoiding the risks happened on patients and the beliefs in services provided by traditional healers. According to [30], quality control is a phrase that refers to processes involved in maintaining the quality or validity of the manufactured products. However, the quality control of herbal medicine is beyond this, meaning it is the management of medicinal plants and their products during cultivation, identification process of the plant species with their parts and localities (their being free from polluted environment causing diseases), and medicine preparation including its components, medication processes, storage standards, and dosage; all should be taken into account. This means, without proper all-round quality control, there is no assurance that the contents of the herbs contained in the package are the same as what are stated outside the package [30]. Climatic factors (prevailing temperature, rainfall, humidity, altitude of the growing region, light), nutritional factors (nutrients, pH, cation exchange capacity), harvesting factors (age, season, collection time, plant organ), and post-harvesting factors (storage hygiene, drying process) are the major factors affecting the contents and composition of medicinal plant raw materials and their products [29, 30]. For these, some of the most important laboratory test methods (histological techniques), common sense, and good pharmaceutical practices are used [29]. Techniques such as thin-layer chromatography and microscopic and electrophoretic techniques are widely used to evaluate the quality of herbal drugs [14, 29, 31] and the content and quality of meats [32] as well. These techniques and good pharmaceutical practices are also used to support the development of national standards based on local market conditions, with due regard to existing national legislation and national and regional norms [29]. Therefore, improved and currently available pharmaceutical analytical methods led to improvements in harvesting schedules, cultivation techniques, storage, product purity, and activity and stability of active compounds [30].

4.2 Major quality control methods for medicinal plant materials and their products

Among others, thin-layer chromatography, macroscopic and microscopic examinations, gas chromatography and volatile components, and electrophoretic techniques [14, 29] are the most important quality control methods for medicinal plant materials and their products, described briefly here below.

4.2.1 Macroscopic and microscopic examinations

Herbal materials are categorized based on sensory, macroscopic, and microscopic characteristics, which are the first steps toward establishing the identity and the degree of purity of such materials, and should be carried out before any further tests undertaken, according to [29]. Therefore, to establish identity, purity, and quality, visual inspection (macroscopic examination) provides the simplest and quickest means. Herbal materials should be entirely free from visible signs of contamination such as insects, molds (fungi), and other animal contamination, including animal excreta; any soil, stones, sand, dust, and other foreign inorganic matter must also be removed before herbal materials are cut or ground for testing [29]. Moreover, plant parts used for medication with abnormal odor, discoloration, slime, or signs of deterioration should be detected to exclude them from being used for medication products.

Moreover, during storage, products should be kept in a clean and hygienic place for avoiding contamination occurring; special care should also be taken to avoid formation of molds, since they may produce aflatoxins [29]. For determination of foreign matter and storage conditions, macroscopic examination can properly be employed for determining the presence of foreign matter in whole or cut plant materials. For these, common sense and good pharmaceutical practices are used. Such common senses and good pharmaceutical practices can, even, be used after laboratory tests since the test procedures cannot take account of all possible impurities in deciding whether an unusual substance not detectable by the prescribed tests can be tolerated [29]. For instance, if a sample is found to be significantly different from the specifications in terms of color, consistency, odor, or taste, it is considered as not fulfilling the requirements. However, such examination may need further microscopic examination for either rejecting or accepting their requirements.

4.2.2 Thin-layer chromatography (TLC)

This technique is simple, can be employed for multiple sample analysis, and so has manyfold possibilities of detection in analyzing herbal medicines [14]. The report of [29] also confirmed that TLC is used for evaluating herbal materials and their preparations; particularly, it is valuable for the qualitative determination of small amounts of impurities.

4.2.3 Gas chromatography (GC) and volatile components

Many pharmacologically active components in herbal medicines are volatile chemical compounds; thereby, the analysis of volatile compounds by gas chromatography is very important in the analysis of herbal medicines [14]. GC is a useful analytical tool in the research field of herbal medicines via analyzing their volatile oils, which have a number of advantages: (1) the GC of the volatile oil gives a reasonable "fingerprint" which can be used to identify the plant and to detect the presence of impurities in the volatile oil, and (2) the extraction of the volatile oil is relatively straightforward and can be standardized, and the components can be readily identified using GC analysis [14].

4.2.4 Electrophoretic method

It is a good tool for producing the chemical fingerprints of the herbal medicines and has similar technical characteristics of liquid chromatography [14]. Electrophoretic method, especially capillary electrophoresis (CE), used in the analysis of herbal medicines, is a versatile and powerful separation tool with a high-separation efficiency and selectivity when analyzing mixtures of low-molecular-mass components [14].

5. Conclusions

There are various forms of medicinal plants including trees, shrubs, climbers, and herbs; of those herbal medicinal plants are dominantly used for different human and animal treatments in Ethiopia. These plants are collected mainly from riverbanks, cultivated areas, bushlands, forest, woodlands, and grasslands, among others. They are used for treatments of stomachaches, dysentery, diarrhea, asthma, cancer, evil eyes, earaches, sores of throat and gum, cough, and so on. For such treatments, these medicinal plants have specific parts used for treatment; most of them are leaves and roots. Hence, traditional medicine plays a significant role in the healthcare of the majority of the people in developing countries, including Ethiopia, and medicinal plants provide valuable contribution to this practice. However, the vegetative resources that are unique to the country, particularly used for medication, are dwindling due to continuous exploitation and pressure on the limited resources. Hence, conservation priority should be given to such medicinal plants and their habitats besides the knowledge of traditional practice of medication via designing appropriate strategies, particularly in the rural areas of the country, where there are less accessibility to clinics and hospitals with their medicines and health experts (doctors). Community- and research-based conservation mechanisms could be an appropriate approach for mitigating the problems pertinent to the loss of medicinal plants and their habitats and for documenting medicinal plants and the knowledge of traditional healers on how to prepare and provide the traditional medication to their patients. Medicinal plants should be multiplied through medicinal gardens, proper handling practices, and scientific development. Moreover, for controlling the quality of medicinal plant materials and their products, chromatography, electrophoretic, macroscopic/microscopic techniques, and pharmaceutical practices are the most important tools.

Acknowledgements

The authors would like to thank Debre Birhan University of Ethiopia for its library facilitation while writing this manuscript. We also extend our thanks to Hirut Fisiha for assisting us during editing and revising of this manuscript.

Conflict of interest

The authors declare that there is no any conflict of interest between authors and other organizations as well.

Author details

Admasu Moges[1*] and Yohannes Moges[2]

1 Department of Biology, College of Natural and Computational Sciences, Debre Birhan University, Debre Birhan, Ethiopia

2 Department of Midwifery, College of Health Sciences, Debre Birhan University, Debre Birhan, Ethiopia

*Address all correspondence to: mogesadmasu@gmail.com

References

[1] Russell SJ, Karunaratne NS, Mahindapala RT. Rapid inventory of wild medicinal plant populations in Sri Lanka. Journal of Biological Conservation. 2006;**132**(4):22-32

[2] Phondani PC, Bhatt ID, Negi VS, Kothyari BP, Batt A, Maikhuri RK. Promoting medicinal plants cultivation as a tool for biodiversity conservation and livelihood enhancement in Indian Himalaya. Journal of Asia-Pacific Biodiversity. 2016;**9**(1):39-46

[3] WHO (World Health Organization). Fact Sheet. Family, Body, Sexuality and Health. Geneva: World Health Organization; 2003. pp. 205-219. Available from: http://books.google.com.et/books.pdf [Accessed: 11 March 2019]

[4] Abebe D, Ayehu A. Medicinal Plants and Enigmatic Health Practices of North Ethiopia. Addis Ababa, Ethiopia: B.S.P.E; 1993

[5] CBD (Convention on Biological Diversity). Forest and Aquatic Plants Genetic Resources. Addis Ababa: Institute of Biodiversity Conservation; 2008

[6] Birhanu T, Abera D, Ejeta E. Ethnobotanical study of medicinal plants in selected HorroGudurru Woredas, Western Ethiopia. Journal of Biology, Agriculture and Healthcare. 2015;**5**(1):83-93

[7] Bekele G, Reddy PR. Ethnobotanical study of medicinal plants used to treat human aliment by Guji Oromo tribes in Abaya district, Borena, Oromia, Ethiopia. University Journal of Plant Science. 2015;**3**(1):1-8

[8] Kelbessa E, Demissew S, Woldu Z, Edwards S. Some threatened endemic plants of Ethiopia. In: Edwards S, Zemede A, editors. The Status of Some Plants in Parts of Tropical Africa. NAPRECA, No. 2. East and Central Africa: Botany 2000; 1992. pp. 35-55

[9] Asfaw Z. The role of home garden in production and conservation of medicinal plants. In: Zewdu M, Demissie A, editors. Conservation and Sustainable Use of Medicinal plants in Ethiopia. Proceeding of the National workshop on Biodiversity Conservation and Sustainable use of medicinal plants in Ethiopia; 28 April–01 May 1998. Addis Ababa: IBCR; 2001. pp. 76-91

[10] Demissie A. Biodiversity conservation of medicinal plants: Problems and prospects. In: Zewdu M, Demissie A, editors. Conservation and Sustainable Use of Medicinal Plants in Ethiopia. Proceeding of the National Workshop on Biodiversity Conservation and Sustainable use of medicinal plants in Ethiopia, 28 April–01 May 1998. Addis Ababa: IBCR; 2001. pp. 56-64

[11] Moges A, Beyene1 A, Triest L, Ambelu A, Kelbessa E. Imbalance of ecosystem services of wetlands and the perception of the local community towards their restoration and management in Jimma Highlands, Southwestern Ethiopia. Wetlands. Available from: https://doi.org/10.1007/s13157-016-0743.pdf [Accessed: 02 February 2019]

[12] Tolossa T, Megersa M. Ethnobotanical Study of Medicinal Plants Used to Treat Human Diseases in Berbere District, Bale Zone of Oromia Regional State, South East Ethiopia. London: Hindawi; 2018. Available from: https://doi.org/10.1155/2018/8602945.pdf [Accessed: 05 February 2019]

[13] Birhane E, Aynekulu E, Mekuria W, Endale D. Management, use and ecology of medicinal plants in the degraded dry lands of Tigray, Northern Ethiopia.

Journal of Medicinal Plant Research. 2011;**5**(3):309-318. Available from

[14] Lianga YZ, Xieb P, Chanc K. Quality control of herbal medicines (Review). Journal of Chromatography B. 2004;**812**:53-70

[15] Tesemama AB, Birnie A, Tengnas B. Useful trees and shrubs for Ethiopia: identification, propagation and management for agricultural and pastoral communities (Technical Handbook No.5). Regional Soil Conservation Unit, Swedish International Development Authority; 1993

[16] Deffar G. Non-wood forest products in Ethiopia. EC-FAO Partnership Program (1998-2000), December, Addis Ababa; 1998

[17] Lulekal E, Kelbessa E, Bekele T, Yineger H. An ethnobotanical study of medicinal plants in Mana Angetu District, Southeastern Ethiopia. Journal of Ethnobiology and Ethnomedicine. 2008. Available from: http://doi.org/10.1186/1746-4269-4-10.pdf [Accessed: 03 February 2019]

[18] Amenu E. Use and management of medicinal plants by indigenous people of Ejaji area (chelya woreda) West shoa, Ethiopia: An ethnobotanical approach [MSc thesis]. Addis Ababa University; 2007

[19] Awas T. Plant diversity in Western Ethiopia: Ecology, ethnobotany and conservation [thesis]. University of Oslo Norway; 2007

[20] Teklehaymanot T. Ethnobotanical study of knowledge and medicinal plants use by the people in Dek Island in Ethiopia. Journal of Ethnopharmacology. 2009;**124**(1):69-78

[21] Yadav RH. Medicinal plants in folk medicine system of Ethiopia. Journal of Poisonous and Medicinal Plants Research. 2013;**1**(1):007-011. ISSN: 2315-8834© 2013. Apex Journal International. Available from: http://www.apexjournal.org.pdf [Accessed: 03 February 2019]

[22] Abera B. Medicinal plants used in traditional medicine by Oromo people, Ghimbi District, Southwest Ethiopia. Journal of Ethnobiology and Ethnomedicine. 2014;**10**(40):1-15. Available from: http://www.ethnobiomed.com/content/10/1/40.pdf [Accessed: 02 February 2019]

[23] Birhanu Z. Traditional use of medicinal plants by the ethnic groups of Gonder Zuria district, North-Western Ethiopia. Journal of Natural Remedies. 2013;**13**(1):46-53

[24] Abebe W. An overview of Ethiopian traditional medicinal plants used for cancer treatment. European Journal of Medicinal Plants. 2016;**14**(4):1-16

[25] Getnet Z, Chandrodyam S, Masresha G. Studies on traditional medicinal plants in ambagiorgis area of Wogera district, Amhara regional state, Ethiopia. International Journal of Pure and Applied Bioscience. 2016;**4**(2):38-45

[26] Doffana ZD. Sacred natural sites, herbal medicine, medicinal plants and their conservation in Sidama, Ethiopia. Cogent Food and Agriculture. 2017;**3**:1365399. DOI: 10.1080/23311932.2017.1365399 [Accessed: 05 February 2019]

[27] Wondimu T, Asfaw Z, Kelbessa E. Ethnobotanical study of medicinal plants around 'Dheeraa' town, Arsi Zone, Ethiopia. Journal of Ethnopharmacology. 2007;**112**(1):152-161

[28] Cunningham AB. Wild plant use and resource management. In: Bennun LA, Aman RA, Crafter SA, editors. The Center for Biodiversity. Nairobi, Kenya: National Museums of Kenya; 1992. pp. 109-126

[29] WHO (World Health Organization). Quality Control Methods for Herbal Materials (Updated edition of Quality Control Methods for Medicinal Plant Materials published in 1998). Malta: WHO; 2011. 187 p. ISBN: 978 92 4 150073 9

[30] El-Alskary H. Quality control of herbal drugs: Total quality management of herbs and medicinal plants, and their products. In: Presentation on International Conference of Ministry of Agriculture and Land Reclamation; 6-8 December; 2011 [Accessed: 02 March 2019]. Available from: https://www.researchgate.net/publication/320306420.pdf [Accessed: 08 March 2019]

[31] Rajput R. Evaluation of Herbal Bioresources Using Histological and Histochemical Techniques. 2000. Available from: pdf.semanticscholar.org [Accessed: 14 March 2019]

[32] Guelmamene R, Bennoune O, Elgroud R. Histological techniques for quality control of meat and meat products—A mini-review. Journal of Nutrition and Human Health. 2018;**2**(2):24-29